Bioprocess Engineering

Bioprocess Engineering

Downstream Processing

Edited by

Pau Loke Show
Chien Wei Ooi
Tau Chuan Ling

CRC Press
Taylor & Francis Group
Boca Raton London New York

CRC Press is an imprint of the
Taylor & Francis Group, an **informa** business

CRC Press
Taylor & Francis Group
6000 Broken Sound Parkway NW, Suite 300
Boca Raton, FL 33487-2742

First issued in paperback 2021

ISBN-13: 978-0-367-77965-8 (pbk)
ISBN-13: 978-1-138-60575-6 (hbk)

Library of Congress Cataloging-in-Publication Data

Names: Show, Pau Loke, editor. | Ooi, Chien Wei, editor. | Ling, Tau Chuan, editor.
Title: Bioprocess Engineering : downstream processing / [edited by] Pau Loke Show, Chien Wei Ooi, Tau Chuan Ling.
Other titles: Bioprocess engineering (Taylor & Francis)
Description: Boca Raton, FL : Taylor & Francis Group, 2019. |
Includes bibliographical references and index.
Identifiers: LCCN 2019002796 | ISBN 9781138605756 (hardback : alk. paper) | ISBN 9780429466731 (ebook)
Subjects: LCSH: Biotechnology. | Biochemical engineering.
Classification: LCC TP248.3 .B5835 2019 | DDC 660.6--dc23
LC record available at https://lccn.loc.gov/2019002796

Visit the Taylor & Francis Web site at
http://www.taylorandfrancis.com

and the CRC Press Web site at
http://www.crcpress.com

Contents

Preface

Downstream processing is a specialized field in bioprocess engineering. It focuses on the isolation and purification of the target product. The biochemical process industry today is growing rapidly due to the advancements in the biotechnological field and the discoveries of novel biologics and therapeutic proteins. Downstream processing is often associated with its counterpart in bioprocess streams: upstream processing. Upstream processing often received greater attention from the biochemical process industry for products such as enzymes, pharmaceuticals, diagnostics, and food. The improvement in upstream processing strategies, for example, the fermentation process, put direct downward pressure on the capability of downstream processing technology, which is often overlooked by industry players. The quality of the final product depends significantly on the performance of downstream processing. Depending on the nature and complexity of processing crude (e.g., fermentation broth, natural sources derived from animal or plant sources, or the waste-associated feed stock), different types of downstream technologies can be adopted.

Bioprocess Engineering: Downstream Processing aims to introduce the commonly used technologies for downstream processing of bio-based products. Topics included in this book are centrifugation, filtration, membrane separation, reverse osmosis, chromatography, biosorption, liquid-liquid separation, and drying. The basic principles and mechanism of separation are covered in each topic, and the engineering concept and design are emphasized. In addition, the latest developments and future prospects of these separation tools are discussed. Chapter 1 is dedicated to the techniques used in cell disruption, which is a prerequisite step for the release of bioproducts prior to the further treatment in the downstream processes.

We aim to present this information as a guide to bioprocess engineers and others who wish to perform downstream processing for their feed stock.

Last, we are grateful to the contributors of the chapters in this book.

Acknowledgments

The editors would like to express sincere gratitude to Dr. Chew Kit Wayne for proofreading and compiling all the chapters in *Bioprocess Engineering: Downstream Processing*.

Editors

Dr. Pau Loke Show is an Associate Professor in the Department of Chemical and Environmental Engineering, Faculty of Engineering, The University of Nottingham Malaysia (UNM). He currently is a Professional Engineer registered with the Board of Engineer Malaysia (BEM) and Chartered Engineer of the Engineering Council UK. He is also a member of the Institution of Chemical Engineers (MIChemE) UK and currently serves as an active member in IChemE Biochemical Engineering Special Interest Group. Dr. Show obtained the Post Graduate Certificate of Higher Education (PGCHE) in 2014, a professional recognition in higher education and is now a fellow of the Higher Education Academy UK (FHEA). He is a world-leading researcher in bioseparation engineering. He establishes international collaboration with researchers from various countries in Asia, Europe, the United States, and Africa. Since he started his career in UNM, he has received numerous prestigious domestic and international academic awards, including The DaSilva Award 2018, JSPS Fellowship 2018, Top 100 Asian Scientists 2017, Asia's Rising Scientists Award 2017, and Winner of Young Researcher in IChemE Malaysia Award 2016. He is currently supervising 10 PhD students. He has published more than 120 journal papers in less than six years of his career. He is now serving as an editorial board member at *Biochemical Engineering Journal* and *BMC Energy*, and is a guest editor for three SCI-indexed journals: *Clean Technologies & Environmental Policy*, *Frontier in Chemistry*, and *BMC Energy*.

Dr. Chien Wei Ooi is an Associate Professor in Chemical Engineering, Monash University Malaysia. He holds a PhD in bioprocess engineering at University Putra Malaysia. His research focuses on the applications of smart polymers and ionic liquids in chemical, biochemical, and bioprocess engineering. He has been involved in research related to bioprocess design and the practical applications of various separation techniques for bioprocess integration. He has secured multiple government research grants and led a number of projects sponsored by chemical and bioprocess industries. He was also engaged as a visiting researcher in universities in Taiwan, Germany, and Portugal.

Dr. Tau Chuan Ling is a Professor of Biotechnology in the Institute of Biological Sciences, University of Malaya. He earned a PhD in Chemical Engineering, Birmingham University, UK; MSc in Environmental Engineering, Universiti Putra Malaysia (UPM); and BSc (Hons) in Biotechnology, Universiti Pertanian Malaysia. He has published more than 150 research articles in a wide range of scientific journals, as well as review articles and patents, in the fields of downstream processing and bioprocess engineering. He has received major grants from the university, the Ministry of Science and Technology of Malaysia (MOSTI), and the Ministry of Higher Education Malaysia (MOHE), and his work has also been funded directly by industry. He has served as visiting professor at Xiamen university (2011) and Tsinghua university (2018). Currently, he is the Co-editor-in-Chief for *Current Biochemical Engineering*.

Contributors

Senthil Kumar Arumugasamy
Department of Chemical and
Environmental Engineering
Faculty of Engineering
University of Nottingham Malaysia
Campus
Semenyih, Malaysia

Jiang Chier Bong
Department of Chemical and
Environmental Engineering
Faculty of Engineering
University of Nottingham Malaysia
Campus
Semenyih, Malaysia

Kit Wayne Chew
Department of Chemical and
Environmental Engineering
Faculty of Engineering
University of Nottingham Malaysia
Campus
Semenyih, Malaysia

Sho Yin Chew
Department of Chemical and
Environmental Engineering
Faculty of Engineering
University of Nottingham Malaysia
Campus
Semenyih, Malaysia

Shir Reen Chia
Department of Chemical and
Environmental Engineering
Faculty of Engineering
University of Nottingham Malaysia
Campus
Semenyih, Malaysia

Suyin Gan
Department of Chemical and
Environmental Engineering
Faculty of Engineering
University of Nottingham Malaysia
Campus
Semenyih, Malaysia

Billie Yan Zhang Hiew
Department of Chemical and
Environmental Engineering
Faculty of Engineering
University of Nottingham Malaysia
Campus
Semenyih, Malaysia

Xing Hu
State Key Laboratory of Food Science
and Technology
School of Food Science and Technology
Nanchang University
Nanchang, China

Kevin Qi Chong Hwang
Department of Chemical and
Environmental Engineering
Faculty of Engineering
University of Nottingham Malaysia
Campus
Semenyih, Malaysia

K. Vogisha Kunjunee
Department of Chemical and
Environmental Engineering
Faculty of Engineering
University of Nottingham Malaysia
Campus
Semenyih, Malaysia

Winn Sen Lam
Department of Chemical and Environmental Engineering
Faculty of Engineering
University of Nottingham Malaysia Campus
Semenyih, Malaysia

Lai Yee Lee
Department of Chemical and Environmental Engineering
Faculty of Engineering
University of Nottingham Malaysia Campus
Semenyih, Malaysia

Hui Yi Leong
Department of Chemical and Environmental Engineering
Faculty of Engineering
University of Nottingham Malaysia Campus
Semenyih, Malaysia

Shuk Yin Lu
Department of Chemical and Environmental Engineering
Faculty of Engineering
University of Nottingham Malaysia Campus
Semenyih, Malaysia

Zahra Motavasel
Department of Chemical and Environmental Engineering
Faculty of Engineering
University of Nottingham Malaysia Campus
Semenyih, Malaysia

Kirupa Sankar Muthuvelu
Department of Biotechnology
Bannari Amman Institute of Technology
Erode, India

Qi Wye Neoh
Department of Chemical and Environmental Engineering
Faculty of Engineering
University of Nottingham Malaysia Campus
Semenyih, Malaysia

Yoong Xin Pang
Department of Chemical and Environmental Engineering
Faculty of Engineering
University of Nottingham Malaysia Campus
Semenyih, Malaysia

Navin Raj Vijiaretnam
Department of Chemical and Environmental Engineering
Faculty of Engineering
University of Nottingham Malaysia Campus
Semenyih, Malaysia

Subbarayalu Ramalakshmi
Department of Chemical Engineering
Dr. MGR Education and Research Institute
Chennai, India

Wei Hon Seah
Department of Chemical and Environmental Engineering
Faculty of Engineering
University of Nottingham Malaysia Campus
Semenyih, Malaysia

Pau Loke Show
Department of Chemical and Environmental Engineering
Faculty of Engineering
University of Nottingham Malaysia Campus
Semenyih, Malaysia

Bervyn Qin Chyuan Tan
Department of Chemical and Environmental Engineering
Faculty of Engineering
University of Nottingham Malaysia Campus
Semenyih, Malaysia

Chung Hong Tan
Department of Chemical and Environmental Engineering
Faculty of Engineering
University of Nottingham Malaysia Campus
Semenyih, Malaysia

Payal Sunil Thadani
Department of Chemical and Environmental Engineering
Faculty of Engineering
University of Nottingham Malaysia Campus
Semenyih, Malaysia

Suchithra Thangalazhy-Gopakumar
Department of Chemical and Environmental Engineering
Faculty of Engineering
University of Nottingham Malaysia Campus
Semenyih, Malaysia

Kai Ling Yu
Institute of Biological Sciences
Faculty of Science
University of Malaya
Kuala Lumpur, Malaysia

Peng Zhang
State Key Laboratory of Food Science and Technology
School of Food Science and Technology
Nanchang University
Nanchang, China

1 Cell Disruption

Subbarayalu Ramalakshmi

1.1 INTRODUCTION

The product secreted during the process of fermentation is either intracellular, extracellular, or periplasmic. If the product is produced extracellularly, the desired product can be obtained from the liquid broth followed by further purification steps. On the other hand, if the product of interest is produced inside the cell (either cytoplasm or periplasm), it is indispensable to disrupt or disturb (in the case of periplasmic expression) the cell in order to extract the intracellular products. Cell disruption involves the pervasion or lysis of the cell that enhances the release of intracellular products (Harrison, 1991). Cells are highly robust in nature. The cell wall provides elasticity and mechanical strength to the cell and therefore provides resistance in order to withstand any sudden pressure changes in the external environment (Booth et al., 2007). Cellular disruption is not a separate procedure but depends on the upstream processes and considerably influences the downstream processes. Downstream operations account for about 70% of the total processing costs (Balasundaram et al., 2009). This requires an appropriate selection of the cell disruption method, which in turn affects the purification steps in downstream operations. Application of the disruption method depends highly on the nature and type of the cell. In case of bacterial cells, they are either gram negative or gram positive, and they differ in the composition of polysaccharide and peptidoglycan (Dmitriev et al., 2005; Silhavy et al., 2010). Yeast cells have a highly complicated extracellular organelle; they are also a rich source of various bioactive compounds, which require careful selection of the disruption method (Lipke and Ovalle, 1998; Okada et al., 2016). A mild disruption method is preferred for mammalian cells, which are easy to disrupt compared to plant cells and microalgae. Other factors such as low cost, maximum product release, ease extraction from the cell debris, and product stability govern the selection of disruption techniques. Cellular disruption methods are briefly categorized into mechanical and nonmechanical methods. Figure 1.1 shows the various existing cell disruption methods for the release of biological products (Harrison, 1991).

Certain measures should be taken before the cell disruption technique in order to increase the efficiency of the disintegration method. After the removal of cells from the culture medium, the products secreted extracellularly and also the media components should be reduced (Balasundaram et al., 2009; Harrison, 1991; Middelberg, 1995).

1.2 MECHANICAL METHODS

Mechanical cell disruption methods use both solid and liquid shear, and the cells are subjected to high pressure/agitation. Chemical agents or external reagents are not added in mechanical cell disruption, which is a chief advantage. Solid shear mainly

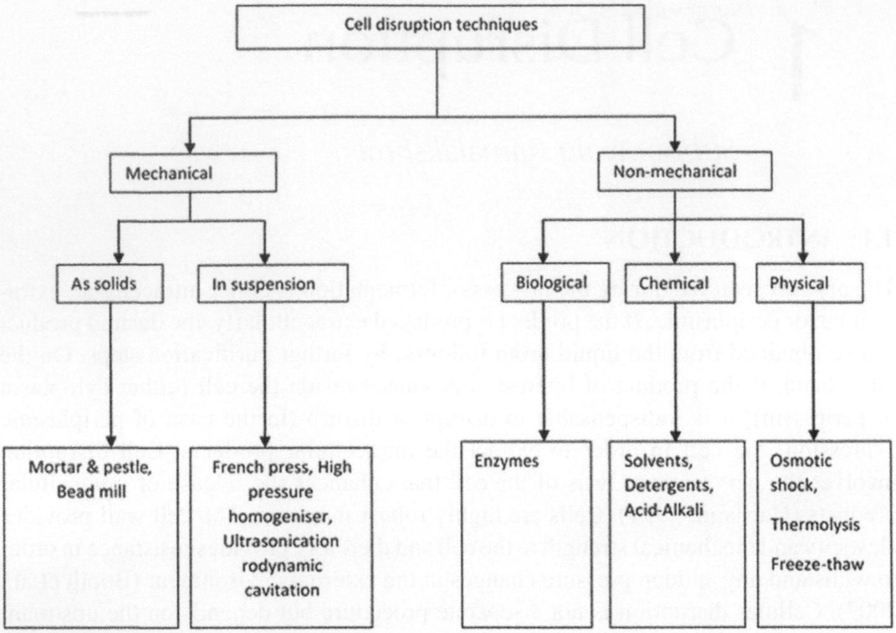

FIGURE 1.1 Different cell disruption techniques. (From Harrison, S.T., *Biotechnol. Adv.*, 9, 217–240, 1991.)

relies on grinding and abrasion, while fluid shear relies on high pressure and velocity of fluid. Pestle-mortar was once a common method to disintegrate cells mechanically and is now rarely used. French press, ultrasonication, and glass bead disruption technology are currently used for small-scale purposes; high-pressure homogenizers (HPHs) and bead mills are widely used in for large-scale purposes. Mechanical methods of disruption have high efficiency and suits almost all types of cells. Most of the methods require cooling after the disruption processes. Heat generation, product degradation, and high cost are some of the drawbacks of mechanical lysis techniques (Goldberg, 2008; Lin and Cai, 2009).

1.2.1 High Pressure Homogenizer (HPH)

HPH is the extensively used method for lysing cells mechanically. Figure 1.2 shows the image of HPH used in industries (Middelberg et al., 1991). Though HPHs are commonly used for large-scale purposes, they are also available on smaller scales and can process 25–200 mL of sample volume (Lin and Cai, 2009). Cells present in the media after harvest are allowed to pass through an orifice (0.1–0.2 mm) where they are compressed by the application of high shear force. High shear, impact, and cavitation are the principles used in HPH. Two storage tanks are available in HPH that work alternatively and process the homogenate. Compressed cells are collected using a positive displacement pump.

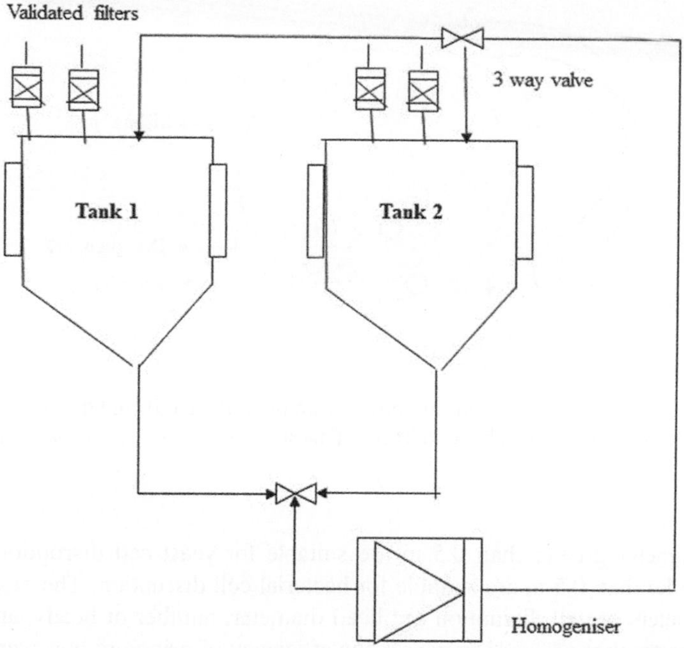

FIGURE 1.2 High-pressure homogenizer used in industries. Tanks 1 and 2 process the cell suspension alternatively, and the cells are disrupted at the valve seats. (From Middelberg, A.P., *Downstream Processing of Proteins*, Springer, pp. 11–21, 2000.)

The amount of protein released by HPH is given by Equation (1.1):

$$\text{In}\frac{R_m}{R_{m-R}} = KNP^a \tag{1.1}$$

where R is protein released; R_m is maximum protein available; P is pressure in MPa; N is the number of passes; K is the rate constant; and a is the pressure exponent (Augenstein et al., 1974; Middelberg, 2000; Middelberg et al., 1991).

High heat evolution and high energy consumption are common problems encountered in this method. An alternative method uses simultaneous emulsification and mixing in order to broaden the usage of HPH (Gall et al., 2016).

1.2.2 Bead Mill

Bead mills are commonly used on a large scale, yet some are also employed in laboratories for disrupting cells. In this technique, cell suspension is mixed with glass, steel, or ceramic beads and agitated at high speed. High shear force is applied to the cells when they collide with the beads, which in turn disintegrate the cell membrane (Figure 1.3) (Pazesh et al., 2017). The type, size, and weight of the beads to be employed largely depends on the nature of the cells to be disrupted. Glass beads

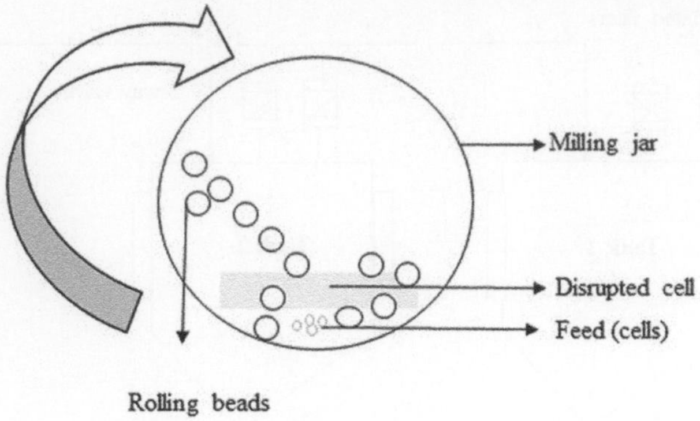

FIGURE 1.3 Bead milling. Beads are mixed with the cells (feed), and the cells are lysed by the principle of grinding and abrasion. (From Pazesh, S., et al., *Int. J. Pharm.*, 528, 215–227, 2017.)

with a diameter greater than 0.5 m are suitable for yeast cell disruption, whereas those smaller than 0.5 m are suitable for bacterial cell disruption. The main governing parameters of cell disruption are bead diameter, number of beads, and agitator speed. The number of beads increases the efficiency of grinding; however, there are problems such as heat evolution, which causes product degradation. Bead shaking is applicable only at lab scale, and bead agitation is often used in industries (Chisti and Moo-Young, 1986; Harrison, 1991; Taskova et al., 2006). Shaking the glass beads with cells can disintegrate more than 60 *E. coli* samples in less than 30 min (Ramanan et al., 2008). Chances of contamination are encountered with this method, which is the main disadvantage.

1.2.3 Ultrasonication and Cavitation

The process of cavitation uses the principle of sonochemistry. In this process, sound energy is generated electrically at a frequency ranging between 20 and 50 Hz. The sound energy travels through a probe that passes through the media solution or water placed in an ultrasonic bath. This process causes the formation of bubbles, which ultimately causes the cell membrane to rupture (Figure 1.4) (Bari et al., 2016). The technique has been used in various cells and in *E.coli* and *Pichia pastoris* (yeast) cells; the cells are processed in less than 0.3 and 1 sec, respectively. The overall rise in temperature during the process is 3.3°C, which is far better than electrical and thermolysis processes that causes product denature. The major challenge is that the process has to be optimized every time depending on the cell type and power input to the device (Tandiono et al., 2012).

Alternate methods are nitrogen cavitation and hydrodynamic cavitation. The physical stress is less in cavitation methods compared to ultrasonic method. Cells are placed in a pressure vessel and nitrogen free of oxygen is passed into the cells

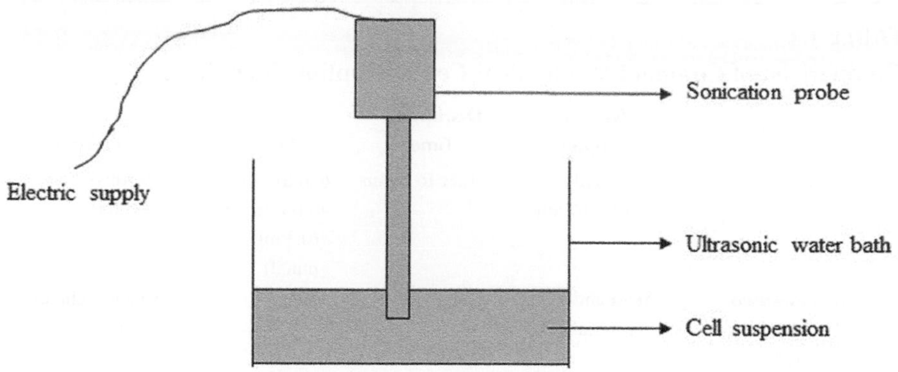

FIGURE 1.4 Ultrasonication process. A sonication probe is used to produce sonic waves for lysing the cells. (From Bari, S., et al., *Ultrason. Sonochem.*, 31, 39–50, 2016.)

under high pressure (approximately 5500 KPa). Nitrogen bubbles are created, which causes the rupture of cell walls. It is best suited for fragile cell walls such as mammalian and plant cells and some bacterial cells, but not for fungi and yeast cells (Simpson, 2010). Hydrodynamic cavitation is an efficient method for the extraction of lipids in microalgae; it also causes less stress on the proteins and enzymes compared to the ultrasonic method. In hydrodynamic cavitation, the sample is passed through a small channel, which increases the velocity, thus causing the membrane to rupture and releasing the intracellular products (Save et al., 1997).

1.2.4 French Press

This is similar to HPH where application of high-pressure technique is employed. Yet this technique is used only for small-scale purposes (Goldberg, 2008). Initially, the cells are passed through a valve into a pump cylinder, after which they are allowed to pass through an annular gap. At this region, the pressure applied is 1500 bar. Then the cells are passed through a discharge valve where the pressure is close to atmospheric pressure. Cell disruption occurs at the discharge valve, where the pressure drops suddenly. This is one of the most prevalent methods used for yeast cell disruption. It is also used in the dairy industry for milk homogenization. Complete disintegration requires more than one pass. However, the number of passes can be reduced by increasing the pressure, which sometimes causes product degradation. Apart from cell concentration and pressure application, the cellular product released depends on the valve and valve seats. The minimum volume of the sample used is 10 mL and the flow rate is approximately 2–3 mL/min. Heat generated during the process can be reduced by external cooling (Geciova et al., 2002; Middelberg, 1995; Pöhland, 1992). Compact mechanical devices are available that use the similar principles of shear, friction, grinding, and abrasion (Table 1.1).

TABLE 1.1
Comparison of Compact Mechanical Cell Disruption Techniques

Devices	Type of Samples	Disruption Time	Volume	References
Ultrasonication	Yeast and bacterial cells	10 sec to 2 min	600 μL^{-1} mL; 40 mL in flow (0.3 mL min^{-1})	Borthwick et al. (2005)
Low-pressure-based micro/nanoscale cell disruption device	Yeast and bacterial cells	20 sec	1 mL	Li and Scherer (2016)
Bead shaking based compact microfluidic device	Gram positive bacteria	3 min	1 mL	Hwang et al. (2011)
Grinding homogenization-based micro/nanoscale cell disruption device	Yeast and bacterial cells	8 min	<70 μL	Kido et al. (2007)
Abrasion-based compact cell disruption device using nano-blade	Mammalian cells	1 min/100 μL	1 to 100 μL min^{-1}	Yun et al. (2010)
Abrasion-based compact cell disruption device using nano-wires	Mammalian cells	1 min/μL	5 μL min^{-1}	Kim et al. (2012)

1.3 PHYSICAL METHODS

Nonmechanical or physical methods of cell disruption do not involve any force to disintegrate the cell. Nonmechanical methods are usually preferred when there is a small sample size and also if there is a need to disrupt any specific part of the cell without any contaminants or a minimal amount of contaminants. Also, they do not cause much shear to the cell, unlike mechanical methods of cell disruption. Hence, physical methods can be applied for cells that do not have a tough cell wall. Sometimes they are combined with mechanical cell disruption methods to achieve complete disintegration of cells (Goldberg, 2008; Shehadul Islam et al., 2017). Freeze-thaw, thermolysis, and osmotic shock are types of physical methods of cell disruption.

1.3.1 Freeze-Thaw

The freeze-thaw technique is used to disrupt mostly mammalian and bacterial cells. It involves submerging the sample cell solution in dry ice or ethanol for 2 min followed by thawing the cells in a water bath at 37°C for about 8 min. This causes the formation of ice on the cells, which leads to rupture of the cell membrane and release of the intracellular components. This method cannot be used for cells that are sensitive to temperature. However, this method is most suitable for highly expressed proteins from *E. coli* and also for isolation of recombinant proteins from cytoplasm. Fifty percent of the recombinant proteins are found to be released in relatively pure form using the freeze-thaw technique (Johnson and Hecht, 1994; Wanarska et al., 2007). Also, this method is proven to be the best technique for the extraction of polysomes from *E. coli*. While other physical and chemical methods failed to remove polysomes, this method involves two cycles of freezing and thawing in the presence of lysozyme (Ron et al., 1966). For the isolation of yeast DNA, the freeze-thaw method proved to be an inexpensive technique compared to enzymatic and glass beads method of disruption (Harju et al., 2004).

However, there are some disadvantages. The freeze-thaw method is time consuming because more cycles of freezing and thawing are required for efficient disruption of cell membrane and release of cellular components. Also, for temperature-sensitive components, this affects the activity. Hence, the process is often combined with other methods of cell disruption to increase efficiency (Part of Thermo Fisher Scientific, version 2, 2009).

1.3.2 Thermolysis

Thermolysis is a simple technique that employs only a stirring tank where the cell suspension is placed. It is an economical method of cell disruption provided the cells are thermally stable (Koschorreck et al., 2017). The principle of heat shock is applied, where the cells are heated to 50°C (periplasmic protein extraction) to disintegrate the cell membrane, leaving the products intact. For the extraction of cytoplasmic proteins, cells have to be heated to 90°C, at which some of the protein molecules and enzymes are unstable and they are degraded. Also, the solubility of the protein is also varied at higher temperatures, which is a factor to consider (Middelberg, 1995). These are the major drawbacks of this method. On the other hand, hyperthermophiles that grow at a temperature of 80°C–110°C can be processed using this method. Enzymes from such organisms, for example, recombinant hyperthermophilic esterase in *E. coli* can withstand high temperatures and they are extracted at 80°C (Ren et al., 2007). Proteases are denatured at such high temperatures, which is an advantage. Bacterial laccases can withstand high temperatures, and this method is best applied for recombinant products (Koschorreck et al., 2017).

1.3.3 Osmotic Shock

Turgor pressure (ranging from 2–6 atm) is necessary for the cells to remain intact and is directly related to the elasticity of cell membranes (Ruiz et al., 2006). However, if the environmental conditions are altered, this pressure varies, which affects the cell's elasticity and in turn the size of pores present in the cell membrane. Variation in pore size causes the release of intracellular contents. In this process, cells are first exposed to hypertonic solution (salt or sugar solution) for them to shrink; then the cells are treated using hypotonic solution (cold water) for the shrunk cells to swell. The shrinking and swelling of cells during hypertonic and hypotonic treatments, respectively, affects the intactness of the cell membrane, thus increasing the pore size. Increase in pore size causes the release of periplasmic proteins, leaving the cytoplasmic contents intact. This is the major advantage of

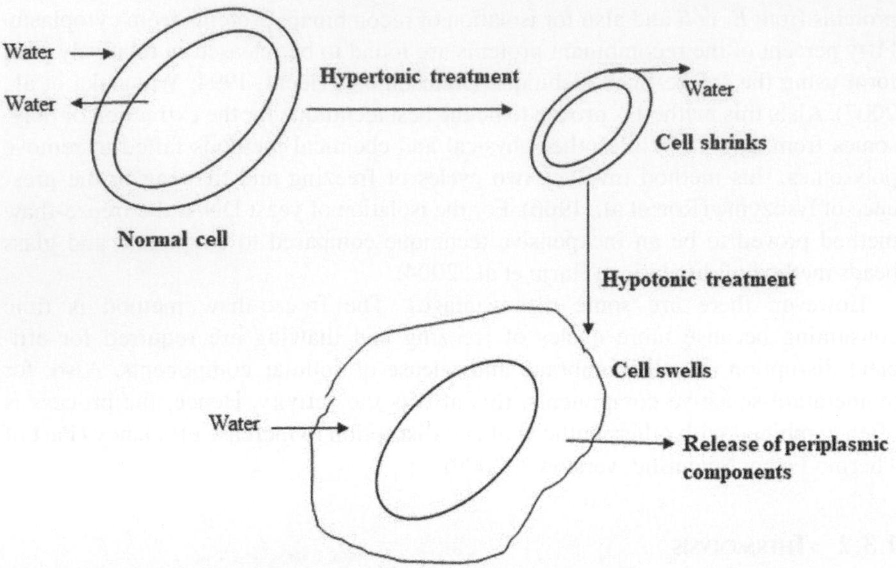

FIGURE 1.5 Osmotic shock process.

this method. It is more suitable for gram negative bacteria and also mammalian cells. The method's efficiency is reduced, however, for cells that have tougher cell walls. Sometimes enzymatic treatment methods are initially applied followed by osmotic shock process, which increases the efficiency of the disruption process (Ramanan et al., 2010; Shehadul Islam et al., 2017). The process of osmotic shock is illustrated in Figure 1.5.

1.4 CHEMICAL AND ENZYMATIC METHODS

In addition to mechanical and physical methods, chemicals and enzymes are also used for cell disruption. Multiple processes are often combined to achieve efficient cell disruption. The main disadvantage of the method is the need to remove the chemical or enzyme after the disruption process to follow the downstream processing steps more easily (Harrison, 1991; Middelberg, 1995). Detergents, solvents, and enzymes are the agents used in chemical and enzymatic methods.

1.4.1 Detergents

Detergents are amphipathic molecules because they contain a polar head and a nonpolar tail. They are used to disrupt the cells or sometimes only the cell membranes for protein extraction. They operate by incorporating themselves into the cell wall or cell membrane, thereby solubilizing lipids and proteins on the cell wall and creating pores on the cell wall. This mechanism results in the release of cellular components such as RNA, DNA, and proteins. A typical cell contains

TABLE 1.2
Comparison of Various Detergents and Their Properties

Detergent	Type	Properties
Sodium dodecyl sulfate (SDS)	Anion (denaturing)	Break protein-protein interactions. Denature protein and destroy protein activity.
Triton X series (114, 100), NP-40, Tween-20, 80	Non-ion (non-denaturing)	Mild in activity. Used when protein activity has to be retained. Suitable for membrane protein extraction.
CHAPS	Zwitterion (non-denaturing)	Mild in activity. Used in enzyme and immunoassays, protein isolation.

Source: A review on macroscale and microscale cell lysis methods, Shehadul Islam, M. et al., *Micromachines*, 8, 83, 2017.

both hydrophilic and hydrophobic molecules. The unique property of detergents is that they disturb the hydrophilic-hydrophobic bonding in cellular components such as lipids, proteins, and polysaccharides. This interaction of detergents with the cell components is based on the charges carried by them. Accordingly, they are classified into anionic, cationic, and non-ionic detergents. The types of detergents and their properties are presented in Table 1.2 (Crapo et al., 2011; Part of Thermo Fisher Scientific, version 2, 2009). Though anionic and cationic detergents work in less time, they cause protein denature. Hence, among the types of detergents, non-ionic detergents are usually used for cell disruption because they have a lesser effect on the cell components such as proteins and enzymes. Non-ionic detergents include Triton series (Koley and Bard, 2010) and zwitterions (3-[(3-cholamidopropyl)dimethylammonio]-1-propanesulfonate (CHAPS). Some of the widely used non-ionic detergents include Triton X-100 that affects the permeability property of the cell in order to extract the proteins. Yet when the detergent exceeds the critical micelle concentration (CMC) value, it causes cell death. This happens when the polar head of the detergent molecule disrupts the hydrogen bonding present in the lipid structure, thus disturbing the integrity of the lipid layer and also causing protein denature. This accounts for the main disadvantage of using detergents in the cell disruption process. However, they are used in minimal amounts for laboratory purposes to extract DNA, RNA, and proteins. In addition, detergents sometimes contain impurities such as peroxides, which affects the activity of protein structures. Also, detergents should be removed from the sample before further downstream operations. Dialysis is a common method for the removal of detergents if they have larger values of CMC. For lower values of CMC, the sucrose gradient method and ion exchange chromatography are used (Harrison, 1991; Scientific; Sharma et al., 2012).

1.4.2 Solvents

Cell disruption, or disturbing the cell membrane to extract proteins by using chemical solvents, is one of the methods commonly used in labs. Solvents such as toluene, alcohol, dimethyl sulfoxide, and methyl ethyl ketone are employed in cell disruption. Cell lysis and the extraction of the components is based directly on the polarity of the solvents. Different solvents suit different types of cell. For the disruption of yeast cells, eight solvents were compared, and it was found that 50% ethanol has the maximum efficiency to permeate the cell membrane (Panesar et al., 2007). Concentration and polarity of the solvents are important parameters to consider. Solvents work by disturbing the lipid membrane in the cell wall, which is compromised to release the intracellular components. In the case of plant cells, solvents are combined with shear forces for the disruption processes.

The major drawback in utilizing solvents is that they damage the cellular components, especially protein molecules. Hence, they have to be used in minimal quantities. Also, some of the solvents cause serious environmental problems. Scaling up is difficult, and the solvent is required in larger amounts when it is applied to biomass. These reasons have restricted the use of solvents to only small-scale processes. Combinations of solvents such as methanol and chloroform give higher yield compared to the use of a single solvent (Byreddy et al., 2015; Stanbury et al., 2013).

Alkaline compounds are also used in the cell lysis process. The working principle is based on the OH^- ion that disrupts the fatty acid-glycerol ester bonds in the cell wall, thus increasing cell permeability for the release of intracellular components. The process is slow, however, and may take up to 12 hours. Alkaline compounds are used along with sodium dodecyl sulfate (SDS), which quickens the process by solubilizing the membrane and proteins. Stability of the product in alkali has to be checked, and the product has to withstand a pH of 10.5–12.5 (Bimboim and Doly, 1979; Klintschar and Neuhuber, 2000).

1.4.3 Enzymes

Different types of enzymes are used to digest cell walls depending on the type of cell wall. Lysozyme is a well-known enzyme used for cell disruption. The enzyme reacts with the peptidoglycan layer directly by hydrolysing the beta 1–4 glycosidic bonds in peptidoglycan. But in the case of gram negative cells, it requires the removal of the outer membrane before treatment with any enzyme. Hence, treatment using enzymes for cell disruption is not very suitable. Zymolyase, chitinase, cellulase, and pectinase are the enzymes used for yeast cell disruption. The addition of enzymes does not cause degradation of proteins or enzymes during extraction. However, enzymes are limited in availability and also very expensive. Hence, their usage is restricted only to labs (Harrison, 1991; Salazar and Asenjo, 2007).

Recently, microfluidics platforms are being used extensively used in the process of cell lysis, and they are classified into mechanical, thermal, chemical, electrical, and acoustic lyses methods (Table 1.3). Single-cell lysis is also a newly evolved field, and the lysis methods use mostly detergents, lasers, and electrical sources (Brown and Audet, 2008).

TABLE 1.3
Different Types of Microfluidic Cell Disruption Techniques

Method	Disruption Efficiency	Features
Mechanical	Medium	It can be applied to any type of cell; expensive (Cheng et al., 2017).
Thermal	High	Easy integration; rapid cell lysis; high heat generation; affects product stability (Packard et al., 2013).
Chemical	High	Rapid lysis; expensive chemical agents (Seo and Yoo, 2018).
Electrical	High	Rapid lysis; easy integration; expensive (Wang et al., 2007).
Acoustic	Medium	Easy fabrication of microfluidics device and electrodes; expensive; rapid cell lysis; heat generation is encountered (Tandiono et al., 2012).

1.5 CONCLUSION

In conclusion, cell disruption methods are vital before any purification step in downstream operations. Mechanical methods are efficient, but they cause product denature in certain cases. Physical methods such as osmotic shock and thermolysis are suitable for only certain types of cells. Chemical methods require the removal of chemicals and enzymes before the purification step. In most cases, a combination of methods results in better yield. Contamination from the disruptor device has to be avoided, and this has a major impact on the quality of the extraction product. Activity of the product is also an important parameter. Cost and speed of the disruption process has to be considered. Microscale methods such as introduction of microfluidics in cell disruption employ the similar macroscale principles in microscale processes. Single-cell lysis methods and selective product recovery are potential areas for future research.

REFERENCES

Augenstein, D., Thrasher, K., Sinskey, A., and Wang, D. (1974). Optimization in the recovery of a labile intracellular enzyme. *Biotechnology and Bioengineering*, 16(11), 1433–1447.

Balasundaram, B., Harrison, S., and Bracewell, D. G. (2009). Advances in product release strategies and impact on bioprocess design. *Trends in Biotechnology*, 27(8), 477–485.

Bari, S., Chatterjee, A., and Mishra, S. (2016). Ultrasonication assisted and surfactant mediated synergistic approach for synthesis of calcium sulfate nano-dendrites. *Ultrasonics Sonochemistry*, 31, 39–50.

Bimboim, H., and Doly, J. (1979). A rapid alkaline extraction procedure for screening recombinant plasmid DNA. *Nucleic Acids Research*, 7(6), 1513–1523.

Booth, I. R., Edwards, M. D., Black, S., Schumann, U., and Miller, S. (2007). Mechanosensitive channels in bacteria: Signs of closure? *Nature Reviews Microbiology*, 5(6), 431.

Borthwick, K., Coakley, W., McDonnell, M., Nowotny, H., Benes, E., and Gröschl, M. (2005). Development of a novel compact sonicator for cell disruption. *Journal of Microbiological Methods*, 60(2), 207–216.

Brown, R. B., and Audet, J. (2008). Current techniques for single-cell lysis. *Journal of the Royal Society Interface*, 5(suppl 2), S131–S138.

Byreddy, A. R., Gupta, A., Barrow, C. J., and Puri, M. (2015). Comparison of cell disruption methods for improving lipid extraction from thraustochytrid strains. *Marine Drugs*, 13(8), 5111–5127.

Cheng, Y., Wang, Y., Wang, Z., Huang, L., Bi, M., Xu, W., and Ye, X. (2017). A mechanical cell disruption microfluidic platform based on an on-chip micropump. *Biomicrofluidics*, 11(2), 024112.

Chisti, Y., and Moo-Young, M. (1986). Disruption of microbial cells for intracellular products. *Enzyme and Microbial Technology*, 8(4), 194–204.

Crapo, P. M., Gilbert, T. W., and Badylak, S. F. (2011). An overview of tissue and whole organ decellularization processes. *Biomaterials*, 32(12), 3233–3243.

Dmitriev, B., Toukach, F., and Ehlers, S. (2005). Towards a comprehensive view of the bacterial cell wall. *Trends in Microbiology*, 13(12), 569–574.

Gall, V., Runde, M., and Schuchmann, H. (2016). Extending applications of high-pressure homogenization by using simultaneous emulsification and mixing (SEM)—An overview. *Processes*, 4(4), 46.

Geciova, J., Bury, D., and Jelen, P. (2002). Methods for disruption of microbial cells for potential use in the dairy industry—A review. *International Dairy Journal*, 12(6), 541–553.

Goldberg, S. (2008). Mechanical/physical methods of cell disruption and tissue homogenization. *2D Page: Sample Preparation and Fractionation*, pp. 3–22. Springer.

Harju, S., Fedosyuk, H., and Peterson, K. R. (2004). Rapid isolation of yeast genomic DNA: Bust n'Grab. *BMC Biotechnology*, 4(1), 8.

Harrison, S. T. (1991). Bacterial cell disruption: A key unit operation in the recovery of intracellular products. *Biotechnology Advances*, 9(2), 217–240.

Hwang, K.-Y., Kwon, S. H., Jung, S.-O., Lim, H.-K., Jung, W.-J., Park, C.-S., and Huh, N. (2011). Miniaturized bead-beating device to automate full DNA sample preparation processes for Gram-positive bacteria. *Lab on a Chip*, 11(21), 3649–3655.

Johnson, B. H., and Hecht, M. H. (1994). Recombinant proteins can be isolated from *E. coli* cells by repeated cycles of freezing and thawing. *Nature Biotechnology*, 12(12), 1357.

Kido, H., Micic, M., Smith, D., Zoval, J., Norton, J., and Madou, M. (2007). A novel, compact disk-like centrifugal microfluidics system for cell lysis and sample homogenization. *Colloids and Surfaces B: Biointerfaces*, 58(1), 44–51.

Kim, J., Hong, J. W., Kim, D. P., Shin, J. H., and Park, I. (2012). Nanowire-integrated microfluidic devices for facile and reagent-free mechanical cell lysis. *Lab on a Chip*, 12(16), 2914–2921.

Klintschar, M., and Neuhuber, F. (2000). Evaluation of an alkaline lysis method for the extraction of DNA from whole blood and forensic stains for STR analysis. *Journal of Forensic Science*, 45(3), 669–673.

Koley, D., and Bard, A. J. (2010). Triton X-100 concentration effects on membrane permeability of a single HeLa cell by scanning electrochemical microscopy (SECM). *Proceedings of the National Academy of Sciences*, 107(39), 16783–16787.

Koschorreck, K., Wahrendorff, F., Biemann, S., Jesse, A., and Urlacher, V. B. (2017). Cell thermolysis—A simple and fast approach for isolation of bacterial laccases with potential to decolorize industrial dyes. *Process Biochemistry*, 56, 171–176.

Li, Z., and Scherer, A. (2016). Handheld low pressure mechanical cell lysis device with single cell resolution. Google Patents. US Patent No. 9,365,816.

Lin, Z., and Cai, Z. (2009). Cell lysis methods for high-throughput screening or miniaturized assays. *Biotechnology Journal*, 4(2), 210–215.

Lipke, P. N., and Ovalle, R. (1998). Cell wall architecture in yeast: new structure and new challenges. *Journal of Bacteriology*, 180(15), 3735–3740.

Middelberg, A. P. (1995). Process-scale disruption of microorganisms. *Biotechnology Advances*, 13(3), 491–551.

Middelberg, A. P. (2000). 2 microbial cell disruption by high-pressure homogenization *Downstream Processing of Proteins*, pp. 11–21. Totowa, NJ: Springer.

Middelberg, A. P., O'Neill, B. K., L. Bogle, I. D., and Snoswell, M. A. (1991). A novel technique for the measurement of disruption in high-pressure homogenization: Studies on *E. coli* containing recombinant inclusion bodies. *Biotechnology and Bioengineering*, 38(4), 363–370.

Okada, H., Kono, K., Neiman, A. M., and Ohya, Y. (2016). Examination and disruption of the yeast cell wall. *Cold Spring Harbor Protocols*, 2016(8), pdb. top078659.

Packard, M. M., Wheeler, E. K., Alocilja, E. C., and Shusteff, M. (2013). Performance evaluation of fast microfluidic thermal lysis of bacteria for diagnostic sample preparation. *Diagnostics*, 3(1), 105–116.

Panesar, P. S., Panesar, R., Singh, R. S., and Bera, M. B. (2007). Permeabilization of yeast cells with organic solvents for β-galactosidase activity. *Research Journal of Microbiology*, 2(1), 34–41.

Pazesh, S., Gråsjö, J., Berggren, J., and Alderborn, G. (2017). Comminution-amorphisation relationships during ball milling of lactose at different milling conditions. *International Journal of Pharmaceutics*, 528(1–2), 215–227.

Pöhland, H. (1992). *Protein Purification: Design and Scale up of Downstream Processing*. Munich, Germany: Hanser Publishers.

Ramanan, R. N., Ling, T. C., and Ariff, A. B. (2008). The Performance of a glass bead shaking technique for the disruption of *Escherichia coli* cells. *Biotechnology and Bioprocess Engineering*, 13(5), 613–623.

Ramanan, R. N., Tan, J. S., Mohamed, M. S., Ling, T. C., Tey, B. T., and Ariff, A. B. (2010). Optimization of osmotic shock process variables for enhancement of the release of periplasmic interferon-α2b from *Escherichia coli* using response surface method. *Process Biochemistry*, 45(2), 196–202.

Ren, X., Yu, D., Yu, L., Gao, G., Han, S., and Feng, Y. (2007). A new study of cell disruption to release recombinant thermostable enzyme from *Escherichia coli* by thermolysis. *Journal of Biotechnology*, 129(4), 668–673.

Ron, E. Z., Kohler, R. E., and Davis, B. D. (1966). Polysomes extracted from *Escherichia coli* by freeze-thaw-lysozyme lysis. *Science*, 153(3740), 1119–1120.

Ruiz, N., Kahne, D., and Silhavy, T. J. (2006). Advances in understanding bacterial outer-membrane biogenesis. *Nature Reviews Microbiology*, 4(1), 57–66.

Salazar, O., and Asenjo, J. A. (2007). Enzymatic lysis of microbial cells. *Biotechnology Letters*, 29(7), 985–994.

Save, S., Pandit, A., and Joshi, J. (1997). Use of hydrodynamic cavitation for large scale microbial cell disruption. *Food and Bioproducts Processing*, 75(1), 41–49.

Seo, M.-J., and Yoo, J.-C. (2018). Lab-on-a-disc platform for automated chemical cell lysis. *Sensors*, 18(3), 687.

Sharma, R., Dill, B. D., Chourey, K., Shah, M., VerBerkmoes, N. C., and Hettich, R. L. (2012). Coupling a detergent lysis/cleanup methodology with intact protein fractionation for enhanced proteome characterization. *Journal of Proteome Research*, 11(12), 6008–6018.

Shehadul Islam, M., Aryasomayajula, A., and Selvaganapathy, P. (2017). A review on macroscale and microscale cell lysis methods. *Micromachines*, 8(3), 83.

Silhavy, T. J., Kahne, D., and Walker, S. (2010). The bacterial cell envelope. *Cold Spring Harbor Perspectives in Biology*, 2(5), a000414.

Simpson, R. J. (2010). Disruption of cultured cells by nitrogen cavitation. *Cold Spring Harbor Protocols*, 2010(11), pdb. prot5513.

Stanbury, P. F., Whitaker, A., and Hall, S. J. (2013). *Principles of Fermentation Technology*. Amsterdam, the Netherlands: Elsevier.

Tandiono, T., Ohl, S.-W., Chin, C. S.-H., Ow, D. S.-W., and Ohl, C.-D. (2012). Cell lysis using acoustic cavitation bubbles in microfluidics. *The Journal of the Acoustical Society of America*, 132(3), 1953–1953.

Taskova, R. M., Zorn, H., Krings, U., Bouws, H., and Berger, R. G. (2006). A comparison of cell wall disruption techniques for the isolation of intracellular metabolites from *Pleurotus* and *Lepista sp.Zeitschrift für Naturforschung C*, 61(5–6), 347–350.

Thermo Fisher Scientific. (2009). *Thermo Scientific Pierce Cell Lysis Technical Handbook*, pp. 2–5, Part of Thermo Fisher Scientific, Version 2.

Thermo Fisher Scientific. (2009). *Thermo Scientific Pierce Cell Lysis Technical Handbook*, p. 5, Part of Thermo Fisher Scientific, Version 2.

Wanarska, M., Hildebrandt, P., and Kur, J. (2007). A freeze-thaw method for disintegration of *Escherichia coli* cells producing T7 lysozyme used in pBAD expression systems. *Acta Biochmica Polonica-English Edition*, 54(3), 671.

Wang, H.-Y., Banada, P. P., Bhunia, A. K., and Lu, C. (2007). Rapid electrical lysis of bacterial cells in a microfluidic device. *Microchip-Based Assay Systems* (pp. 23–35): Springer.

Yun, S.-S., Yoon, S. Y., Song, M.-K., Im, S.-H., Kim, S., Lee, J.-H., and Yang, S. (2010). Handheld mechanical cell lysis chip with ultra-sharp silicon nano-blade arrays for rapid intracellular protein extraction. *Lab on a Chip*, 10(11), 1442–1446.

2 Centrifugation

Xing Hu and Peng Zhang

2.1 INTRODUCTION

Centrifugation is a technique that uses the centrifugal force generated by the rotation of a centrifugal head and container in a centrifuge as well as the difference in sedimentation coefficient, size, shape, or buoyant density of the sample material to be separated from suspended solids (Anlauf, 2007). Since the first industrial three-legged centrifuge was introduced in Germany in 1836, the separation technology has been considerably improved. Centrifugal separation can be used not only for the direct recovery of liquid or solid in suspension but also for the separation of two mutually different solutions. Almost every laboratory or industrial production unit involved in bioprocess engineering is equipped with various types of centrifuges. Similar to filtration and membrane separation, centrifugal separation is one of the most important and widely used solid-liquid separation technologies used in biological processes. It is commonly used for the recovery or removal of microorganisms in bioengineering operations; the separation of proteins, enzymes, nucleic acids, viruses, and cellular subcomponents; and the clarification, concentration, classification, and enrichment of liquids (Shukla et al., 2007). Anlauf (2007) reviewed the tasks for centrifugal separation, as shown in Figure 2.1, which can be defined very differently. The process is especially well suited for the separation of molecularly immiscible liquids, which are frequently in the presence of solid particles.

2.2 PRINCIPLES OF CENTRIFUGAL SEDIMENTATION

The principles of centrifugal sedimentation can be analyzed through the gravitational settling of solids (Price, 1982). As shown in Figure 2.2, gravity leads the solid particles to undergo gravitational settling in infinitely continuous fluids, wherein the solid particles are subjected to the joint forces of gravity, buoyancy, and viscous drag.

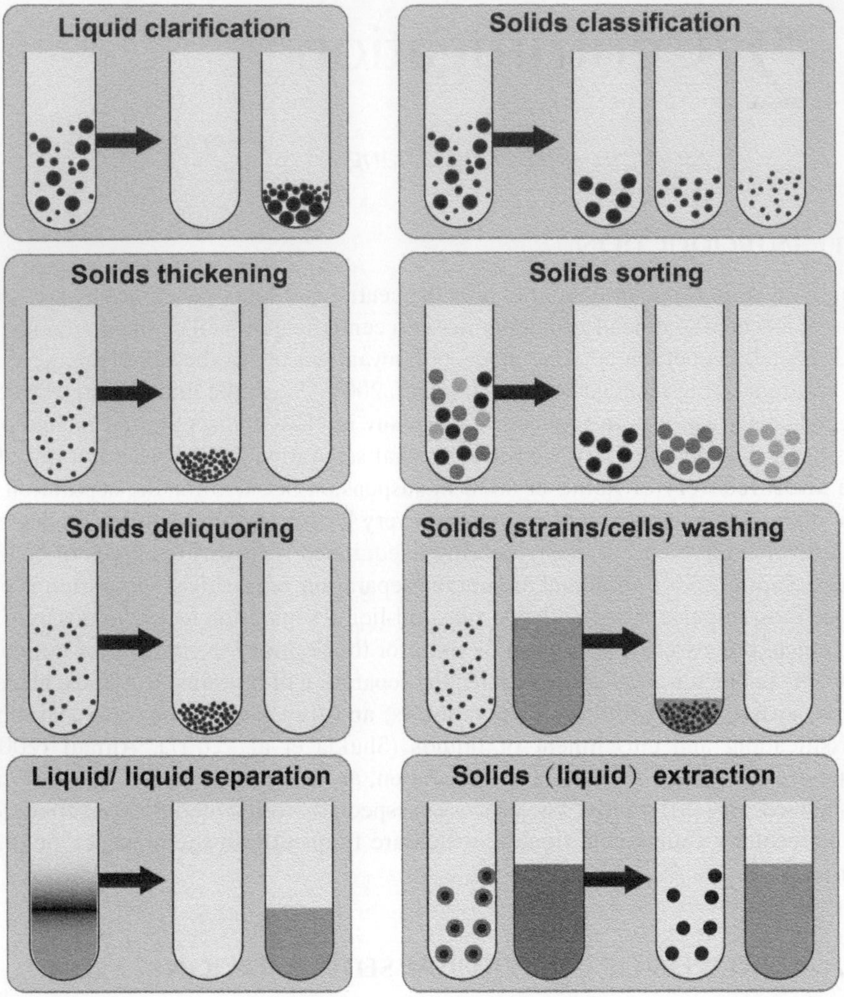

FIGURE 2.1 Tasks for centrifugal separation.

If one considers the object of bioseparation to be a spherical particle with a diameter d_P, the buoyancy force F_B that affects this particle can be expressed by the following equation:

$$F_B = (\rho_S - \rho_L)gV = \frac{\pi d_P^3(\rho_S - \rho_L)g}{6} \tag{2.1}$$

where d_P is the particle diameter (m); ρ_s and ρ_L represent solid density and liquid density, respectively (kg/m^3); g is the gravitational acceleration (m/s^2); and V represents

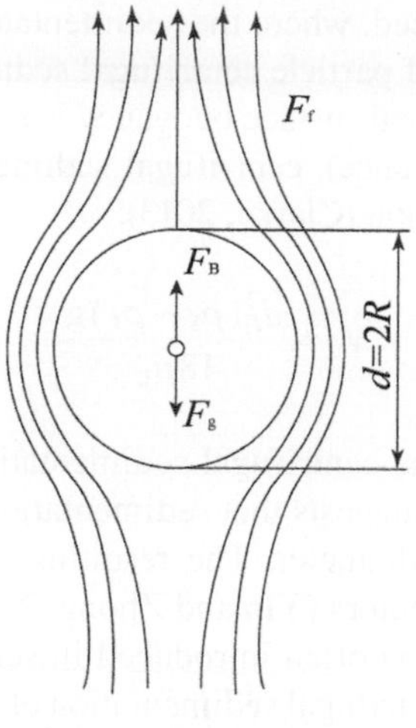

FIGURE 2.2 Forces of spherical particles in settlement process.

particle volume. According to Stokes's theorem, F_f (the viscous drag force of the spherical particles suspended in the medium) can be expressed as:

$$F_f = 3\pi d \mu_L v_g = \frac{1}{2} C_D A \rho_L v_g^2 \tag{2.2}$$

where μ_L represents the viscosity of the continuous fluid, v_g corresponds to particle velocity, C_D is the retardation coefficient, and A is the projected area of the particles in the motion direction.

C_D is not a constant and derives from the change in the Reynolds number R_e. With regard to spherical particles, the experimental evidence suggests that (Shuler and Kargi, 2001):

$$R_e < 1 \qquad C_D = \frac{24}{R_e} \tag{2.3}$$

$$1 < R_e < 10^4 \qquad C_D = \frac{24}{R_e} + \frac{3}{\sqrt{R_e}} + 0.34 \tag{2.4}$$

The Reynolds number R_e is defined as:

$$R_e = \frac{d_p v_g \rho_L}{\mu_L} \tag{2.5}$$

The centrifugation process involved in biological operations is primarily focused on cells. Particle diameter is relatively insignificant and results in a fairly low centrifugal sedimentation speed, where the sedimentation process can meet the criteria of $R_e < 1$. If the speed of particle centrifugal sedimentation reaches a constant value, F_B (that is, if the spherical surface buoyancy is equal to F_f—the force resisting the liquid, or frictional resistance), centrifugal sedimentation velocity is generally determined by Stokes's equation (Clarke, 2013):

$$v_g = \frac{d_P^2(\rho_S - \rho_L)g}{18\mu_L} \tag{2.6}$$

where v_g is the velocity of the centrifugal sedimentation. As for nonspherical particles, the relevant literature suggests that sedimentation velocity should first be calculated with the equivalent diameter. The resulting value should then be refined according to particle-shape factors (Yin and Zhong, 2008).

Centrifugation equipment is often introduced in scientific research and real-life production to speed up the centrifugal sedimentation of heterogeneous phase, thereby mitigating the low efficiency of natural sedimentation. The heterogeneous system is driven by the centrifugal force field and rotates around a certain axis. An inertial centrifugal force is thereby generated. Once particle density exceeds liquid density, solid particles undergo a settling process akin to a gravitational field. Gravity drives this natural settling process, while centrifugal force drives that of centrifugal sedimentation. On these grounds, the centrifugal force received by each mass unit is:

$$F_c = r\omega^2 = 4\pi N^2 r \tag{2.7}$$

where ω, whose unit is rad/s, is the angular velocity at which the centrifugal rotor rotates; r is the centrifugal radius (cm), that is, the distance between the central axis of the rotor to the settled particles.

Compared with centrifugal sedimentation, centrifugation simply transfers the gravitational driving force to the centrifugal driving force. By replacing the gravitational acceleration g with the centrifugal acceleration F_c, the following equation can be obtained:

$$v_s = \frac{d_P^2(\rho_S - \rho_L)r\omega^2}{18\mu_L} \tag{2.8}$$

Equation (2.8) is the fundamental centrifugal sedimentation formula. It demonstrates that the larger the actual centrifugal radius, and the higher the rotational speed of the centrifugal device, the greater the centrifugal force. Therefore, in order to enhance the effect of centrifugal separation, we can strengthen the centrifugal force, increase the centrifugal radius, increase the difference between the particles and the solution, or reduce the viscosity of the suspending medium. In addition, K_c—the centrifugal separation factor—can be used for the following quantitative evaluation:

$$K_c = \frac{F_c}{g} = \frac{r\omega^2}{g} \tag{2.9}$$

K_c stands for the ratio of centrifugal acceleration to particle freefall acceleration in a centrifuge. $K_c = 1000$ implies that the centrifugal force is 1000 times superior to gravity. A larger K_c can improve the separation process.

2.3 TYPES OF CENTRIFUGAL SEPARATIONS

The centrifugal separation method can be roughly classified into differential centrifugation, density gradient centrifugation, and centrifugal elutriation on the basis of the underlying methods (Graham, 2001b).

2.3.1 Differential Centrifugation

Differential centrifugation or pelleting is the simplest and most commonly used method for centrifugal separation of biological entities. It is a commonly employed technique used for separating biologically active substances, such as animal and plant viruses, various subcellular fractions (nuclei, chloroplasts, and mitochondria), as well as crude extraction and concentration of biological macromolecules (nucleic acids and proteins) (Graham, 2001a). Differential centrifugation generally uses fixed-angle rotors to separate particles of different sizes and densities by gradually increasing the centrifugal speed. First, the heaviest particles completely sink to the bottom of the tube during low-speed centrifugation. The supernatant is then precipitated by centrifugation at slightly higher rotational speed to obtain a secondary particle sample. By gradually increasing the centrifugal rotation speed, sample particles of different weights can be obtained to achieve the purpose of separation. During each step of operation, light particles close to the bottom of the tube cause some interference and contaminate the precipitate. To avoid this situation, it is necessary to resuspend the precipitate and repeat this centrifugation step several times to obtain particles of uniform size.

In general, the separation effect of differential centrifugation is positively related to the differences in particle size and density. Table 2.1 shows the sizes of partial bacterial cells and subcellular organelles (Tan, 2007). In practice, operating conditions, including centrifugal rotational speed and time, shall be selected in line with the characteristics, purposes, and extents of the centrifugal separation, thereby effectively separating the various components in the given feed liquid. Ye et al. (2008) used differential centrifugation to obtain single cell suspension in the culture of neural stem cells and obtained satisfactory results. This method can preserve the function of neural stem cells and remove the debris caused by mechanical dissociation. According to Livshits et al. (2016) when it comes to exosomes, under suitable differential centrifugation conditions, the cut-off size of the vesicles can be estimated via a relatively simple theory. In addition, the study developed an online-based interactive centrifugal parameter calculator (http://vesicles.niifhm.ru/), which can be used to simplify the calculation of the relevant parameters for readers. Besides, in some isolation protocols, differential centrifugation combined with other separation methods, such as biomagnetic separation, can result in a more highly enriched target population.

TABLE 2.1
Size of Some Strains and Subcellular Organelles

Particle	*d* (μm)
Nuclei	4–12
Plasma membrane sheets	3–20
Escherichia coli	2–4
Yeast	2–7
Mitochondria	0.4–2.5
Lysosomes	0.4–0.8
Peroxisomes	0.4–0.8
Vesicles	0.05–0.4

2.3.2 Zonal Centrifugation

Zonal centrifugation is an important separation method used in bioengineering. According to different centrifugal operating conditions, it can be divided into rate-zonal and isopycnic density-gradient sedimentation. The commonality between these two methods is as follows: a density gradient medium with certain low molecular weight solutions (such as sucrose, glycerol, KBr, and CsCl) is prepared prior to centrifugation. The medium gradient should be formed first, and the maximum density of the medium should be less than that of all sample particles. Then, the material liquid to be treated is added to the solution with the density gradient to be centrifuged. The principle of rate-zonal density-gradient sedimentation is that different components of the sedimentation coefficient form different zones in the density gradient. During centrifugation, the maximum density in the density gradient is less than the density of the target product to be separated. In the density gradient, each component of the feed liquid settles at different rates due to differences in their sedimentation coefficient, resulting in zones corresponding to the components. After centrifugation for a specific time period, various purified components can be obtained by aspirating the zones formed at different positions in the centrifuge tube. The density gradient of equilibrium zone centrifugation is larger than that of differential zone centrifugation. After centrifugation, the polymer solute in the feed liquid forms a stable zone at the solvent, the density of which is equal to its own density, and the solute concentration in the zone is a Gaussian distribution centered on the density.

Zonal centrifugation is generally suitable for the separation and purification of biological macromolecules such as proteins and nucleic acids; however, it has small processing capacity and can be used only in a laboratory scale. Zonal rotors can be used instead of centrifugal tubes to increase the processing capacity of zonal centrifugation.

Lin et al. (2013) presented a readily scalable purification approach utilizing rate-zonal centrifugation for purifying DNA-origami nanostructures, which provides separation resolution comparable to that of agarose gel electrophoresis. The DNA nanostructures remain in aqueous solution throughout the purification process. Therefore, the desired products are easily recovered with consistently high yield

(40%–80%) and without contaminants such as residual agarose gel or DNA intercalating dyes. Michinaka and Fujii (2012) applied isopycnic centrifugation, and fractionation to extracting DNA samples from batch culture, thus separating fructose fermenters from nonfermenters. Mustroph et al. (2009) described a method for the isolation and quantification of polysome complexes from plant tissues; when centrifugally separating ribosome subunits, ribosomes, and polyribosomes, the detergent-treated cell extracts can be separated using high-speed differential centrifugation and further purified using centrifugation through sucrose density gradients. Other approaches to improve centrifugation efficiency such as the combination use of differential centrifugation and zonal centrifugation have also been reported in the literature.

2.3.3 Centrifugal Elutriation

Cell samples separation is one of the basic preparation steps in conventional biological operations. In the late 1940s, Lindahl designed a counterstreaming centrifuge that could separate cells according to the velocity of cell sedimentation. Later, Beckman Instruments improved the counterstreaming centrifuge and named it elutriator, and the process was called centrifugal elutriation (also known as counterstreaming centrifugation). Centrifugal elutriation is an advanced installation that increases the sedimentation rate to improve the resolution of cell separation. This installation can separate a variety of cells for its sedimentation rate, for example, hemopoietic cells, mouse tumor cells, testicular cells, and so on. This method has a small influence for cell function, but it is not suitable for cell separation of cells with similar sedimentation characteristics and different kinds of cells (Banfalvi, 2011).

Gillespie and Henriques (2006) took centrifugal elutriation as a means of cell cycle phase separation and synchronization: elutriated DT40 cells in growth medium at room temperature using a constant flow rate (40 mL/min) and slowing the rotor to predetermined speeds to elute the desired fraction(s) of cells. They found that the appropriate rotor speed must be determined by experience and can vary according to the cell's genotype. In traditional counterflow centrifugal elutriation, the balance of centrifugal force and flow force can be changed gradually by increasing the velocity of flow or decreasing the speed of rotation so that cell separation and recovery can be realized based on the difference of size, shape, and/or density. Morijiri et al. (2013) reported a microfluidic system that used counterflow centrifugal elutriation, density-gradient media, and the branching inlet structures. Using this system, they successfully separated the microparticles and observed the cells' retention behavior in a separation chamber. There was no need for traditional pump-integrated rotor systems in counterflow centrifugal elutriation by using density-gradient media, which greatly simplifies the instrument and program.

2.4 CENTRIFUGAL EQUIPMENT

A centrifuge is a commonly used piece of separation equipment in the bioengineering sector. Lab centrifuges generally adopt a manual unloading mode and a tubular rotor structure, with an intermittent centrifugal operation (Ghosh, 2006).

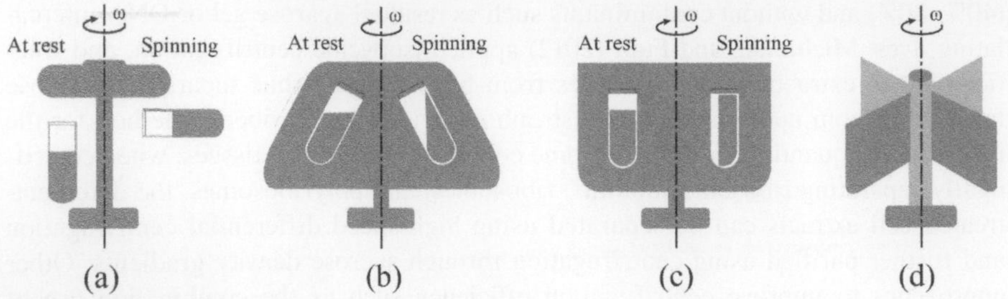

FIGURE 2.3 Different types of centrifuge rotors. (a) Swing-out rotors; (b) Fixed-angle rotors; (c) Vertical rotors; and (d) Zonal rotor.

Figure 2.3 shows various types of centrifuge rotors. Commonly used rotors include swing-out rotors (Figure 2.3a), fixed-angle rotors (Figure 2.3b), vertical rotors (Figure 2.3c), zonal rotor (Figure 2.3d), and so on.

In practice, the relative centrifugal force K_c, also referred to as the relative centrifugal force (RCF) value, is determined by the centrifuge rotational speed and by the centrifuge tube placement in the rotor. Centrifuges are usually classified by centrifugal separation equipment according to their K_c value: if K_c is less than 7000, the centrifuge is identified as a constant speed centrifuge; if K_c is in the 7000 to 21000 range, it's considered a high-speed centrifuge; and if K_c exceeds 21000, it is called an ultracentrifuge. Bioengineering projects generally use refrigerated centrifuges, which can be operated at low temperatures. Industrial centrifugal equipments generally require a large processing capacity and a capacity for continuous operations. A wide range of centrifuges are currently available, such as tubular-bowl centrifuges and disc-bowl centrifuges. The tubular-bowl centrifuge, also known as a cylindrical centrifuge, has a relatively simple structure. In centrifugal operations, liquid is brought in from the center of the round tube. This liquid in the tube is then driven by centrifugal force to form a cylindrical surface centered on the rotating shaft, while the supernatant is discharged from the center of the other end. Solid particles settle on the tube wall during intermittent operations. During continuous operations, the concentrated suspension is discharged from the outlet near the tube wall. The revolution of the tubular-bowl centrifuge can be rather high, and thus the centrifugal force can reach a significant level (Doran, 2012). However, the settlement area (Σ) is small and its treatment capacity is insufficient.

Assuming that r_1, represents the distance from the middle point of the rotation axis to the liquid surface of the tubular-bowl centrifuge, and that the inner diameter of the tubular bowl is r_2, as shown in Figure 2.4a, then the settlement area (Σ) of the tubular-bowl centrifuge can be determined by the following equation:

$$\Sigma = \frac{\pi\omega^2 b}{2g}(3r_2^2 + r_1^2) \tag{2.10}$$

A large number of equally-spaced disc separators are introduced in the centrifugal rotor of the disc-bowl (Figure 2.4b) centrifuge in order to enlarge the settlement area (Subramanian, 1998). This structure offers a strong processing capacity.

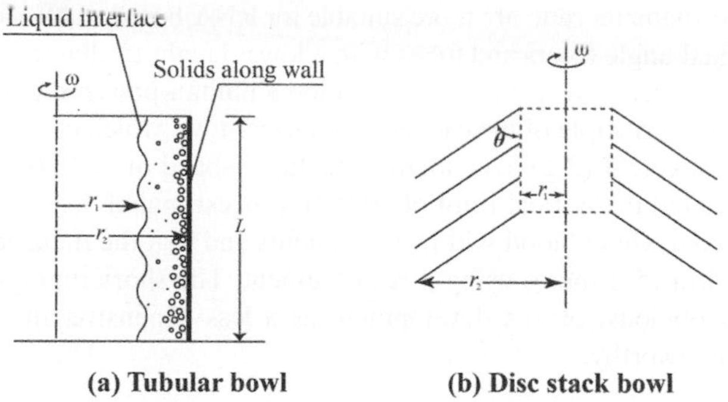

FIGURE 2.4 Types of centrifuge. (a) Tubular-bowl centrifuge and (b) disc-bowl centrifuge.

However, the internal structure of this type of centrifuge is extremely complicated, and the centrifugal revolution is outperformed by tubular-bowl centrifuges. The settlement area of disc-bowl centrifuges is as follows:

$$\Sigma = \frac{2\pi n\omega^2}{3g}(r_2^3 - r_1^3)\cot\theta \tag{2.11}$$

Centrifuge selection ought to preferentially involve those models that meet the parameters to ensure that the centrifuge is equipped with the terminal settling rate and separation capacity required for the separation process. For geometrically and dynamically similar centrifuges, their centrifugal settlement areas Σ can be used as a criterion for a rough comparison. Assuming that the volumetric feeding rates of the two centrifuges are Q_1 and Q_2, respectively, and that their respective sedimentation areas are Σ_1 and Σ_2, then:

$$\frac{Q_1}{\Sigma_1} = \frac{Q_2}{\Sigma_2} \tag{2.12}$$

However, the results derived from this comparison are merely estimates. In practice, empirical evidence is also taken into consideration. A reasonable choice can be made by including both the characteristics of the liquid to be processed and the separation properties of a wide range of centrifuges.

Centrifugal technology, especially ultracentrifugation, has improved gradually with the development of science and technology. The ultracentrifuge has a powerful centrifugal force field and relatively high separation efficiency, and thus it is the most convenient and efficient equipment for separating and purifying biological macromolecules such as viruses, proteins, and plasmid DNA. It can be subdivided into analytical and preparative centrifuges. The analytical ultracentrifuge can be operated at low temperatures to preserve sample activity; the preparative ultracentrifuge has a large load and higher separation efficiency. Herrmann et al. (2017) studied the effect of rotor type on the separation of isotope-labeled and unlabeled *Escherichia coli* RNA by isopycnic density ultracentrifugation. They found that (near-) vertical rotors and tubes with a larger height-to-diameter ratio are more suitable for RNA-based stable isotope probing studies than fixed-angle rotors and tubes with a lower height-to-diameter ratio.

Bhamla et al. (2017) were the first to develop a human-powered paper centrifuge based on the basic principle of the carousel (or buzzer toy), which costs approximately 20 cents, is lightweight (2 g), can attain centrifugal speed of 125,000 (30,000 × *g*), and has a predicted theoretical limit of 10,000. It is estimated that pure plasma can be separated from whole blood within 1.5 minutes and that the malaria parasite can be isolated within 15 minutes using this instrument. The shortcomings of the paper centrifuge are obvious, but its development as a less-expensive alternative of the centrifuge is noteworthy.

2.5 CONCLUSION

Centrifugal separation, which is an indispensable part of bioengineering, has the advantages of high separation speed and efficiency, and good liquid clarity, and it is suitable for large-scale separations; however, the centrifugal equipment is expensive, and the process is energy-intensive. Furthermore, the solid phase drying degree is not equivalent to that of filtration, sterilization is inconvenient, and the process is associated with the risk of aerosol contamination, as microbial aerosols of 10 to 100 CFU/m^3 concentration are generated in case of accidental tube rupture (Jin et al., 2018). High-speed ultracentrifuges may cause serious accidents due to their high speed and large centrifugal force, improper use, or lack of regular inspection and maintenance. Therefore, the operating procedures must be strictly observed. Clark (2001) provides an overview of modern centrifuges in practical aspects, including care and safe use of centrifuges and rotors, as well as prevention of rotor failure and other accidents. Studies have shown that centrifugation does not significantly affect properties of biological samples but may cause cell lysis (Joseph et al., 2016; Streit et al., 2010). Therefore, during biological treatment, the literature should be consulted and practical experience should be considered to select the appropriate centrifuge and centrifugation conditions with the aim of minimizing the effect of centrifugation on the sample. In certain cases, other techniques such as filtration can be used along with different centrifugation modes to reduce expenses and improve efficiency (Felo et al., 2013).

REFERENCES

Anlauf, H. 2007. Recent developments in centrifuge technology. *Separation and Purification Technology* 58(2): 242–246.

Banfalvi, G. 2011. Synchronization of mammalian cells and nuclei by centrifugal elutriation. *Methods of Molecular Biology* 761: 31–52.

Bhamla, M. S., B. Benson, C. Chai, G. Katsikis, A. Johri, and M. Prakash. 2017. Hand-powered ultralow-cost paper centrifuge. *Nature Biomedical Engineering* 1: 0009.

Clark, D. E. 2001. Safety and the laboratory centrifuge. *Chemical Health and Safety* 8(6): 7–13.

Clarke, K. G. 2013. *Bioprocess Engineering: An Introductory Engineering and Life Science Approach*. Woodhead Publishing: Oxford, UK.

Doran, P. M. 2012. *Bioprocess Engineering Principles* (second edition). Academic Press: Waltham, MA.

Felo, M., B. Christensen, and J. Higgins. 2013. Process cost and facility considerations in the selection of primary cell culture clarification technology. *Biotechnology Progress* 29(5): 1239–1245.

Ghosh, R. 2006. *Principles of Bioseparations Engineering*. World Scientific Publishing: Singapore.

Gillespie, D. A., and C. M. Henriques. 2006. Centrifugal elutriation as a means of cell cycle phase separation and synchronisation. *Sub-Cellular Biochemistry* 40: 359–361.

Graham, J. 2001b. *Biological Centrifugation*. London, UK: John Graham Research Consultancy.

Graham, J. M. 2001a. Isolation of mitochondria from tissues and cells by differential centrifugation. *Current Protocols in Cell Biology* pp. 3.3.1–3.3.15.

Herrmann, E., P. Koch, C. U. Riedel, W. Young, and M. Egert. 2017. Effect of rotor type on the separation of isotope-labeled and unlabeled Escherichia coli RNA by isopycnic density ultracentrifugation. *Canadian Journal of Microbiology* 63(1): 83–87.

Jin, A. J., L. F. Hu, K. Zhang et al. 2018. Quantitative analysis of bio-contamination generated during experiment activities and accidents on the indoor environment in BSL-3 laboratory. *Chinese Medicine Biotechnology* 13(2): 97–102.

Joseph, A., B. Kenty, M. Mollet et al. 2016. A scale-down mimic for mapping the process performance of centrifugation, depth and sterile filtration. *Biotechnology and Bioengineering* 113(9): 1934–1941.

Lin, C., S. D. Perrault, M. Kwak, F. Graf, and W. M. Shih. 2013. Purification of DNA-origami nanostructures by rate-zonal centrifugation. *Nucleic Acids Research* 41(2): e40.

Livshits, M. A., E. Khomyakova, E. G. Evtushenko et al. 2016. Isolation of exosomes by differential centrifugation: Theoretical analysis of a commonly used protocol. *Scientific Reports* 6(1): 21447.

Michinaka, A., and T. Fujii. 2012. Efficient and direct identification of fructose fermenting and non-fermenting bacteria from calf gut microbiota using stable isotope probing and modified T-RFLP. *Journal of General and Applied Microbiology* 58(4): 297–307.

Morijiri, T., M. Yamada, T. Hikida, and M. Seki. 2013. Microfluidic counterflow centrifugal elutriation system for sedimentation-based cell separation. *Microfluid Nanofluid* 14(6): 1049–1057.

Mustroph, A., P. Juntawong, and J. Bailey-serres. 2009. Isolation of plant polysomal mRNA by differential centrifugation and ribosome immunopurification methods. *Methods of Molecular Biology* 553: 109–126.

Price, C. A. 1982. *Centrifugation in Density Gradients*. New Brunswick, NJ: Waksman Institute of Microbiology.

Shukla, A. A., M. R. Etzel, and S. Gadam. 2007. *Process Scale Bioseparations for the Biopharmaceutical Industry*. New York: Taylor & Francis Group.

Shuler, M. L., and F. Kargi. 2001. *Bioprocess Engineering Basic Concepts* (second edition). Upper Saddle River, NJ: Prentice Hall.

Streit, F., G. Corrieu, and C. Beal. 2010. Effect of centrifugation conditions on the cryotolerance of *Lactobacillus bulgaricus* CFL1. *Food and Bioprocess Technology* 3(1): 36–42.

Subramanian, G. 1998. *Bioseparation and Bioprocessing: Biochromatography. Membrane Separations, Modeling, Validation*. Weinheim, UK: Wiley-VCH.

Tan, T. W. 2007. *Bioseparation Technology*. Beijing, China: Chemical Industry Publishing House (in Chinese).

Ye, S., Z. Su, J. Zhang, X. Qian, Q. Zhuge, and Y. Zeng. 2008. Differential centrifugation in culture and differentiation of rat neural stem cells. *Cellular and Molecular Neurobiology* 28(4): 511–517.

Yin, F. H., and J. Zhong. 2008. *Modern Separation Technology*. Beijing, China: Chemical Industry Publishing House (in Chinese).

3 Filtration

Shir Reen Chia, Winn Sen Lam,
Wei Hon Seah, and Pau Loke Show

3.1 INTRODUCTION

Filtration is the process of using a filter medium to separate solid matter from fluids. In filtration, a liquid or gaseous fluid containing solid particles passes through a filter with fine pores that allows the fluid to pass through, while the oversized solid particles that are unable to pass through the pores of the filter are retained. The fluid that passes through the filter is called filtrate or permeate; the particles that cannot passed through the filters are retentate. Different types of filter media with various pore sizes are available in the market. Depending on the filtration requirements, membrane filters with smaller pore sizes may use to remove extremely minute solid particles such as proteins and viruses.

Filtration has been used since the early ages of human civilization, most notably in water purification. Slow sand filtration is one of the earliest known examples of large-scale water treatment; it is performed using a sand bed as the filter. The build-up of biological material and microorganisms results in a thin biofilm that acts as a filter in the upper layers of the sand bed that traps and confines bacteria. Slow sand filtration can be performed intermittently or continuously. Both methods are relatively simplistic in design, but, most important, they can produce potable water with minimal residue, lower bacterial content, and better overall water quality (Jenkins et al., 2011). Intermittent slow sand filtration remains an effective option of water treatment for underdeveloped countries because it does not require a continuous water source (Tiwari Sangya-Sangam et al., 2009). In addition, membrane filters are commonly used in industrial water treatment processes to treat leachate that causes soil and ground water pollution (Bennctt, 2015). The advent of modern technology has allowed further development of sophisticated filtration systems to improve filtrate purity and other useful research and industrial applications such as tangential flow filtration (TFF) to isolate proteins or purify viruses (Grzenia et al., 2008; Palmer Andre et al., 2008).

In general, there are two types of filtration techniques: gas filtration and solid-liquid filtration. Gas filtration is very important in order to ensure clean and sterilized air for indoor environments as well as to negate pollution effects to the outdoor environment. A recent review on the development of air and gas filtration technology has shown that gas filtration remains increasingly relevant in numerous industries such as the pharmaceutical, food and beverage, and energy sectors (Bennett, 2016). Solid-liquid filtration is frequently used in many of the same industries, albeit to serve a different purpose of purifying a liquid stream to obtain either a residue-free filtrate or to retain valuable solids (Toledano et al., 2010; Hjorth et al., 2011).

Other than filtration, there are plenty of other separation processes, such as distillation, adsorption, crystallization, and more. The major difference between other separation processes and filtration is the utilization of a membrane that serves as an interface to separate two bulk phases (Scott and Hughes, 2012). One of the conventional material separation processes is distillation; it is particularly useful in separating homogeneous mixtures with components with different boiling points through heating and condensing. Nevertheless, the separation of heterogeneous mixtures such as liquids or gases that contain solids is unable to be performed through distillation. If the objective is to separate solids from a gaseous stream, gas filtration is typically cheaper and much simpler to implement using filters or membranes than using wet scrubbers that rely on residence times and the use of chemical solutions. For solid-liquid filtration, centrifugation is faster and more efficient than filtration, but it is more expensive because trained specialists are needed to operate the centrifuge. However, a study reported by Udén (2006) has shown that centrifugation may not necessarily be better than filtration techniques in the recovery of insoluble fibers (Udén, 2006). The suitability of each method is subjected to certain factors such as performance, cost, efficiency, and simplicity.

The biotechnology and pharmaceutical industries have flourished over the years due to the increasing importance of medicine and sustainable environmental resources. For pharmaceutical companies, the great emphasis on drug safety necessitates the use of filtration in combination with centrifugation to carry out solid recovery as well as to maintain product purity. In addition, the development of superior filter membranes has played an important role in the biotechnology sector by allowing the creation of improved medicine and medical processes such as dialysis through ultrafiltration. Toxic industrial and environmental wastes are also being treated with filtration, thus effectively reducing any potential adverse environmental effects (Sutherland, 2011).

This chapter will discuss two filtration techniques in detail: gas filtration and solid-liquid filtration. Two modes of filtration, dead-end filtration and cross-flow filtration, are also presented in this chapter with their respective advantages and disadvantages. The remaining sections will contain additional information about the widespread application of filtration, recent advancements in filtration techniques, and future potential developments and challenges of filtration technology.

3.2 FILTRATION TECHNIQUES

3.2.1 Theory

3.2.1.1 Gas Filtration

Gas filtration techniques in separating air and gas from solids, including highly hydrophobic membranes, have various applications such as tank, lyophilizer venting, and service gas filtration (Yazgan-Birgi et al., 2018). Air filtration techniques are known as the best possible method because all individual applications usually have a very precise requirement. Sometimes having too much or too little of an ingredient makes a lot of differences in the filtration performances (Liu et al., 2017). An optimized filter that can meet the requirements in different applications requires mixing of different filters with different individual specifications. Hence, some filter cartridges consist of high pleat density to achieve a more effective filtration rate. High pleat density means there is a multitude of membrane folds in the pleated cylinder, making the folds thinner and thus causing capillary action and moisture to be held within the pleats (Rebai et al., 2010). This reduces the total time needed to reach the initial air flow and a longer blow-down time through a filter cartridge.

Air flow is part of the running costs within the facility. Achieving high air flow through the vent filter is necessary to achieve the demanded air flow with low investment costs. Air filters are required to achieve optimized flow rates as the costs in terms of energy consumptions are higher when the differential pressure required to reach the needed air flow rate is high (Gormley et al., 2017). Sizing of the filter is another important factor in ensuring that the filter system is reliable. Improper sizing causes high running costs and higher costs of equipment repair due to damage. Making the air filter smaller for the ventilation of the tank to be cost competitive in the investment phase of the project eventually results in higher running costs in the long term.

3.2.1.2 Solid-Liquid Filtration

In solid-liquid filtrations, the desired scale of particle removal can be attained by using more than one filter; this increases filtration efficiency by having fewer particles to be filtered. Prefiltration is usually applied at an upstream filter. Prefiltration serves to separate the larger particles and contaminants before sending them to the downstream or final filters (Dixit and Braeutigam, 2007). Depth type prefilters are made of fibrous materials with larger pores for the removal of larger particles while allowing smaller particles to pass through. Depth filters follow the laws of chance because they are constructed by progressive deposition of fibers, meaning that, regardless of the attempts in arranging their positions, the fiber placements still take place randomly. Depth filters are mats of larger pore size distribution and can be classified only by a nominal retention rating (Chew and Law, 2018). Repetitive stacking of fibers in the fabrication of depth filters make it thicker than the membranes (Selatile et al., 2018). Depth filters offer greater dirt loading capacities due to the interstices within the fiber matrix, which allows more inner space to accommodate the particle penetrations.

Membrane filtration methods are commonly used in the sterilization of heat-sensitive liquid products (Stanbury et al., 2017). The Food and Drug Administration (FDA) encourages sterilization through heating whenever possible (Tipnis and Burgess, 2017). However, not all products can withstand sterilization through heating because they can be damaged by heating. Protein is an example of a heat-sensitive product that can be denatured in a heating process. That is when sterilization by filtration becomes helpful. This method has widely applied in various industries, especially in food and drug industrial applications in solving the sterilization of heat-labile products. Therefore, the combination of both prefiltration and membrane filtration has enhanced the efficiency of the filtration removal rate of the particulates, leading to better cost savings by just changing cheaper prefilters for the greater expense of the final membrane filters. Placing prefilters in a series before the final downstream filters prolongs the service life of the final filters by sparing them part of the particulate burden (Parker Hannifin plc, 1993).

3.2.2 Mechanism

Both gas and liquid filtrations have the similar filtration mechanisms. These are discussed in the following sections.

3.2.2.1 Direct Interception

Bigger particle sizes than the pore size of the membrane are separated from the stream. Average particle sizes of the hydrophobic membrane filters are usually >0.2 microns (Das and Waychal, 2016). Direct interception is an effective filtration for both liquid and gas services. Any particle sizes larger than the pores of the filter medium are removed by the openings or holes of the filter medium as direct interception, as shown in Figure 3.1 (Yu et al., 2009).

3.2.2.2 Diffusional Interception

The purpose of diffusional interception is to separate particles that are extremely small. Small particles colliding with the liquid molecules causes the particles to move randomly in the fluid flow lines during the process. This movement is known as Brownian motion, and it can be observed as the gas particles collide with filter materials and are trapped as they moved randomly due to low viscosity of the air and collisions with gas molecules for the particles of sizes <0.35 microns (Maddineni et al., 2017). Figure 3.2 shows that the particles stick on the filter media; they will then be dislodged with G forces. The deviation of the particles from the fluid flow lines has increased the chances of the particles sticking to fiber surfaces and being filtered (Maddineni et al., 2017).

3.2.2.3 Inertial Impaction

Inertial impaction is the most effective separation mechanisms for particles of >1 micron with high density and velocity (Maddineni et al., 2017). The bigger particles are not able to follow the air streamline, which suddenly diverts around the filtration media and hits the filter media, as shown in Figure 3.3. This separation mechanism depends on the momentum associated with the particles in the fluid stream. Liquid streams flow through the least resistance and are diverted around the fiber; particles with larger momentum travel in a straight line, causing them to strike on the fiber and be removed (Liu et al., 2016).

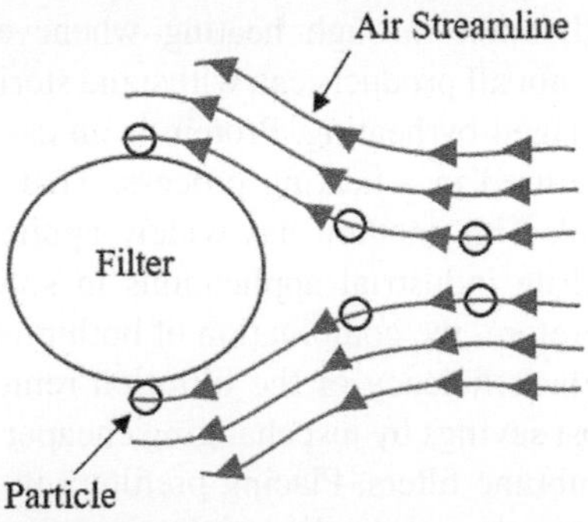

FIGURE 3.1 Direct interception.

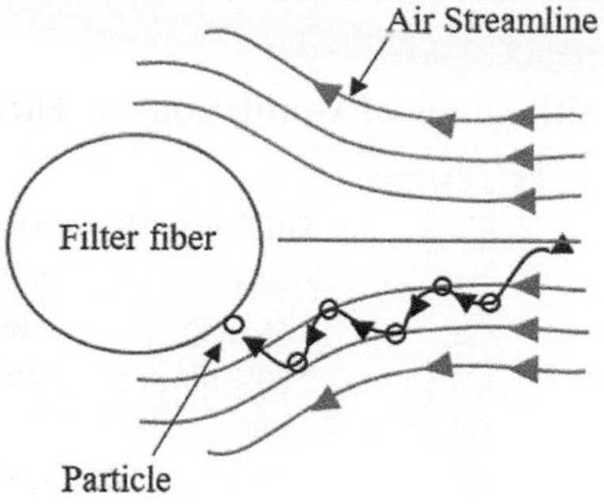

FIGURE 3.2 Diffusional interception.

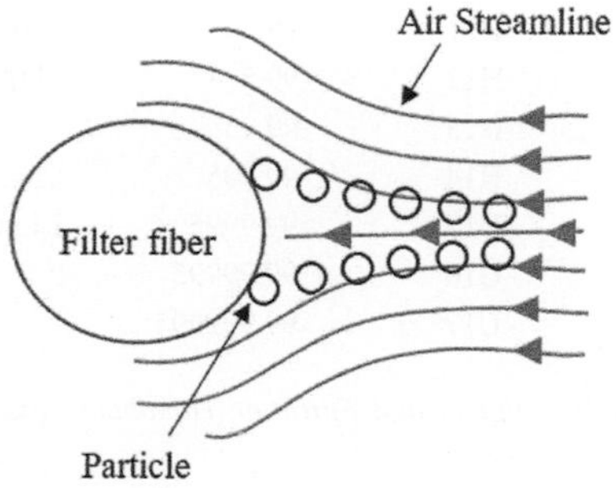

FIGURE 3.3 Inertial impaction.

3.2.3 Filter Media Selection

3.2.3.1 Gas Filtration

The type of filter is one of the crucial elements in filtering gas with impurities or polluted air. The selection of air filter media depends on the degree of protection and volume of air to be treated. Gas filtration is required in domestic applications such as normal room protection and ventilation systems, and it is required to supply sterilized air for industrial applications such as biomedical and critical processes. Filters with the capability of an absolute cut-off are essential to protect against very small particles or submicrometer particles with particle sizes between 5 and 10 μm (Xie et al., 2018); filters with normal cut-off are sufficient for less critical applications.

There are three basic performance characteristics to be evaluated when selecting a suitable filter media: the separation efficiency against the particle size, pressure drop performance, and contaminant holding capacity (Tang et al., 2018). Table 3.1 shows the separation efficiency of several types of filters commonly applied in industrial applications.

TABLE 3.1
Eurovent and CEN Classifications of Ventilation Air Filters

Type of Filters	Eurovent Class	CEN EN779 Class	Efficiency (%)	Measured by	Standards
Coarse dust filter	EU1	G1	<65	Synthetic dust weight arrestance	ASHRAE 52–76 Eurovent 4/5
	EU2	G2	65 < 80		
	EU3	G3	80 < 90		
	EU4	G4	>90		
Fine dust filter	EU5	F5	40 < 60	Atmospheric dust spot efficiency	BS 6540 DIN 24 185 EN 779
	EU6	F6	60 < 80		
	EU7	F7	80 < 90		
	EU8	F8	90 < 95		
	EU9	F9	>95		
High-efficiency particulate air (HEPA) filter	EU10	H10	85	Sodium chloride or liquid aerosol	BS 3928 Eurovent 4/5 DIN 24 184 (DIN 24 183)
	EU11	H11	95		
	EU12	H12	99.5		
	EU13	H13	99.95		
	EU14	H14	99.995		
Ultra-low penetration air (ULPA) filter	EU15	U15	99.9995	Liquid aerosol	DIN 24 184 (DIN 24 183)
	EU16	U16	99.99995		
	EU17	U17	99.999995		

Source: Sparks, T. and Chase, G., *Filters and Filtration Handbook*, Elsevier Science & Technology, Oxford, UK, 2013.

3.2.3.2 Solid-Liquid Filtration

Selection of a liquid filter medium starts with a definition of the filter duty in separation; for example, knowing how fine a particle size can be achieved in the separation process, and the cut-off point and sharpness of the cut-offs is helpful in choosing an appropriate filter medium. Every application requires its own unique requirements, and there is no one solution to all in the standard filter selection procedure. In most cases, there is more than one suitable type of filter available for each set of duty requirements. Table 3.2 summarizes the types of fluid filters.

3.2.4 Applications

3.2.4.1 Gas Filtration

Until the 1900s, there were no recorded instances of filtration being used for any large-scale industrial purpose other than solid-liquid filtration in water treatment. In the case of air or gas filtration, the original high-efficiency particulate air (HEPA) filter was designed and invented in the 1940s during the Manhattan Project in order to limit the dispersion of radioactive contaminants in the air. The HEPA filter was capable of improving air quality greatly because it could filter 99.97% of airborne particulates that are equivalent to or larger than 0.3 μm in size (Gantz, 2012). Since then, the term HEPA has been used to refer to any air filters with high filtration efficiency.

TABLE 3.2
Basic Types of Fluid Filters

Type of Filter	Media	Remarks
Surface	Resin-impregnated paper (usually pleated)	Capable of fine filtering
		Low permeability
	Fine-woven fabric cloth (pleated or "star" form)	Lower resistance than paper
		Ultra-fine filtering
	Membranes	Coarse filtering and straining
	Wire mesh and perforated metal	
Depth	Random fibrous materials	Low resistance and high dirt capacity.
	Felts	Porosity can be controlled/graduated by the manufacturer.
	Sintered elements	Provide both surface and depth filtering. Low resistance.
		Sintered metals mainly but ceramics for high-temperature filters.
Edge	Stacked discs	Paper media are capable of extremely fine filtering.
	Helical wound ribbon	Metallic media have high strength and rigidity.
Precoat	Diatomaceous earth, perlite powdered volcanic rock, and so on	Form filter beds deposited on flexible, semi-flexible, or rigid elements.
		Particularly suitable for liquid clarification.
Absorbent	Activated clays	Effective for removal of some dissolved contaminants in water, oils, and so on.
	Activated charcoal	Also used as precoat or filter bed material.
		Particularly used as drinking water filters.

Source: Sparks, T. and Chase, G., *Filters and Filtration Handbook*, Elsevier Science & Technology, Oxford, UK, 2013.

This specification has become a technical standard for a HEPA filter under the American Society of Mechanical Engineers (ASME) AG-1 code, which is recognized by the United States Department of Energy (2015).

In the *Filters and Filtration Handbook*, it is stated that air filters designed for air treatment can be classified into three categories (Sparks and Chase, 2013):

- *Primary filters*: These air filters are designed with a high dust-holding capacity to filter large dust particles (>5 μm) in the air. They also remain effective at high flow rates of air.
- *Second-stage filters*: These filters have a finer filter medium to retain smaller airborne particles (0.5 to 5 μm) that would normally pass through primary filters. These filters operate at a relatively low air flow of 0.12 m/s or less.
- *Ultra-fine/final-stage filters*: These high-efficiency filters can retain more than 99% of the airborne particles; these are usually HEPA and ultra-low penetration air (ULPA) filters. The maximum air velocities for these filters are limited to very low flow rates (less than 0.03 m/s).

All three categories can be incorporated into a single multistage filtration system to provide high-efficiency air filtration. The filters gradually become obstructed as the dust and particulates continue to build up over time, subsequently causing loss in filtration efficiency and potentially damaging the filters. Therefore, such systems require proper maintenance and replacement of the filter media to preserve filtration performance.

Electrostatic precipitators (ESPs) are a modern gas filtration technology that uses an ionizer to induce electric charge toward dust particles as they pass through the ionizer. The charged dust particles then pass through and are deposited onto several oppositely charged plates called electrodes. ESPs are usually considered secondary filters that work in conjunction with a primary filter that retains larger airborne particles. ESPs have been applied in coal-fired power plants to remove large amounts of air particulates that are generated during power plant operations (Wang et al., 2008; Guo et al., 2014).

Filtration is widely used in our daily lives. Most air filters are used in heating, ventilation and air conditioning (HVAC) systems to control the internal air flow and indoor air quality. Indoor air conditioning not only provides comfort to the occupants of a room by cooling down the room's environment but also purifies the air inside the room. Modern air-conditioning systems contain air filters that are an integral part of removing microorganisms, particulates, and even ozone in the air (Zhao et al., 2007; Yu et al., 2009). A previous study has demonstrated that particle air filtration in air-conditioning systems bring significantly larger benefits than the operating costs involved (Bekö et al., 2008). This is because cleaner indoor air decreases the risk of disease and improves the mortality rates of the occupants.

Modern vehicles employ filters in conjunction with catalytic converters, especially in diesel vehicles, to remove the soot and solid particulates generated by diesel engines (Alkemade and Schumann, 2006). Gas filters are also used to protect the drivers and passengers of vehicles by maintaining clean air flow into the vehicle cabin. Automobiles and trucks take in air from the surrounding areas while filtering out the airborne particulates such as dirt, dust, and soot. Air recirculation systems in large passenger vehicles such as airplanes rely on HEPA filters or adsorption filters (filters that contain adsorbents to bind with gaseous contaminants) to remove odors and airborne microbes such as bacteria and viruses (Bull, 2008).

The use of filtration is commonly found in the pharmaceutical and biotechnology industries due to the extreme importance of having a clean environment with sterile air. Pharmaceutical products such as medicine or supplements are manufactured to treat ailments and improve human health; therefore any contaminants present in the drug or supplement may potentially bring severe adverse effects if consumed. The same applies to the food industry, where sterility is important in preventing pathogens or dust that compromise food quality. Therefore, it is vital to ensure that the production process operates in an environment with sterile air filtered by HEPA filters. In additional, HEPA filters can be used at both entry and exit points to prevent any spread or leakage of airborne material into the exterior environment during manufacturing processes that involve toxic gases or very fine powdered solids (Sutherland, 2011).

3.2.4.2 Solid-Liquid Filtration

Filtration remains one of the most preferred methods in performing the separation of solid-liquid mixtures. Filtration can be used whether the undesired component to be separated in the mixture is a solid from a liquid or a liquid from a solid. In the biotechnology industry, the production of proteins typically involves the cultivation of cells or microorganisms in a bioreactor. As these cells develop and multiply in a growth medium, they simultaneously produce desired proteins. After the growth process has been completed, these proteins need to be purified by separating them from the surrounding liquor that contains the leftover cells and nutrients. This is usually done by implementing centrifugation as an initial step to segregate each component in the mixture (Sutherland, 2011). Alternatively, microfiltration or ultrafiltration can serve as an additional step or employed alone without centrifugation to achieve further purification of the proteins (Ge et al., 2005; Baldasso et al., 2011). In another study, high-performance tangential flow filtration (HPTFF) was shown to be effective in the purification of proteins, specifically antibodies (Saxena et al., 2009). This study demonstrated that HPTFF successfully achieved a 12% increase in yield compared to conventional chromatography techniques because the use of a charged membrane with larger pores in HPTFF allowed for higher selectivity and yield (Lebreton et al., 2008).

Different applications of cross-flow membranes in the food industry remain widely used today (Lipnizki, 2010). Membrane filtration is prevalent in the food industry as it has many advantages, such as high selectivity, low energy consumption, and milder treatment effects, which are suitable for food processing. Raw milk can be microfiltered to remove bacteria and spores, and then further pasteurized into bacteria-free milk or filtered through ultrafiltration to create whey protein concentrates (Baldasso et al., 2011). In addition, ultrafiltration of milk can be used to produce cheese since it has the added benefit of increasing the total solids and therefore the overall cheese yield (Pouliot, 2008). The effluent from the dairy industry can be treated with reverse osmosis to recover water to be reused for heating, cooling, and even cleaning purposes (Vourch et al., 2008).

The brewing industry has also seen large growth over the years as the demand for beer has increased. Advances in brewing technology have allowed brewers all over the world to manufacture high-quality beer at large scale. Filtration is important in ensuring that the undesired solids present during the beer production process can be properly separated from the final product. A study investigating solid wastes in the brewing process mentioned that filtration is used to improve colloidal stability and reduce the cloudiness of the unfiltered beer, which is caused by residual particulates such as proteins, yeast, and certain carbohydrates (Mathias, 2014). It is stated that diatomaceous earth is widely used in the brewing industry as a cost-effective alternative to press filters since diatomaceous earth is exceptionally porous and capable of being used as a depth filter. Conventional dead-end filtration is typically used by many brewers where the spent grain and barley husks form the filter bed on the membrane. However, the need for periodic removal of the spent grain limits it to batch processes. Several studies related to cross-flow filtration in the separation of spent grain and husks have been performed, but it was found that this technique resulted in poor filtration rates. The stability of the membranes is also affected because the husks are abrasive and generated fouling effects (Schneider et al., 2005). Microfiltration is also one of the most commonly used membrane separation methods used in brewing operations since it is applicable to almost every process in production (Ambrosi et al., 2014). Some of these processes include the separation of mash, clarification of beer, cold sterilization, and beer recovery.

On the other hand, the handling of wastewater is very crucial for production plants. If industrial wastewater is not managed properly, there could be long-term community and environmental adverse effects. In water treatment plants, filters are used to remove biological contaminants and other solid wastes. The wastes generated by the pharmaceutical industry are placed under special scrutiny because of the various chemicals involved, such as catalysts, biological compounds, and volatile organic compounds (VOCs) used in tablet coatings (Boltic et al., 2013). Solvent extraction relies on other separation methods such as distillation or evaporation, while filtration is useful in removing solid wastes such as antibiotics and enzymes. Microfiltration, ultrafiltration, nanofiltration, or reverse osmosis may be used for purification depending on the desired purity of the wastewater filtrate or the size of the solids to be removed. It is cheaper to use larger-pore membranes such as microfiltration if extremely high-purity water, such as water produced from reverse osmosis, is not required. Wastewater treatment in the textile industry can be enhanced by implementing microfiltration and nanofiltration as a post-treatment method, after the coagulation-flocculation of the effluent has been performed (Ellouze et al., 2012). A review of water treatment technologies stated that certain membrane processes are more suitable for specific purposes. Microfiltration is effective in removing bacteria and microorganisms; ultrafiltration can be used for macromolecules such as sugars and viruses; and nanofiltration and reverse osmosis can separate divalent ions and monovalent ions, respectively (Gadipelly et al., 2014). For example, nanofiltration is quite effective in separating the antibiotic, amoxicillin, from pharmaceutical wastewater. The effectiveness of nanofiltration not only reduces the costs, it also prevents external pollution and potential damage to the environment (Shahtalebi, 2011). Table 3.3 shows examples of filtration applications and the techniques used in various industries.

TABLE 3.3
Application of Filtration in Various Industries

Industry	Application	Filtration Technique	References
Biotechnology	Protein (e.g., enzymes, antibodies, etc.) purification, cell separation	MF, NF, UF, HPTFF	Saxena et al. (2009)
Pharmaceutical	Solvent recovery, drug recovery	MF, NF, UF, RO	Gadipelly et al. (2014)
Dairy	Milk production, cheese production, protein concentrate production	MF, NF, UF, RO	Baldasso et al. (2011)
Brewing	Mash separation, beer clarification, cold sterilization, beer recovery	MF, NF, UF, RO	Ambrosi et al. (2014)
Water treatment	Solid waste removal, bacterial removal	MF, NF, UF, RO	Ellouze et al. (2012), Vourch et al. (2008)

3.3 FILTRATION MODE

3.3.1 Dead-End Filtration (Conventional Filtration)

Dead-end filtration is the most basic form of filtration, perhaps the oldest of all the practical options available to engineers. This is a conventional filtration method that is often used as the starting point for separation because this technique has been applied for over 100 years and serves as a ready-made reference point. One can assess more modern options to choose a better alternative to conventional filtration (Wu et al., 2018).

Conventional filtration is said to occur when cells and other solids are separated from the liquid broth with the aid of pressure or a vacuum (Mah et al., 2014). The separation is carried out in a filter or a filter press. The typical particle size of the solids removed is in the range of 0.2 μm to several micrometers. Figure 3.4 shows that the flow of liquid to be filtered is directly perpendicular to the filter surface. The complete feed flow is forced through the membrane, and the filtered matter is accumulated on the surface of the membrane.

The performance of filtration is expressed in filtrate flux, J. The definition of filtrate flux is filtrate flow rate per unit of membrane area. It is also related to the pressure through the following expression (Ní Mhurchú and Foley, 2006):

$$J = \frac{\Delta P}{\mu(R_m + \alpha m)} \tag{3.1}$$

where J is the filtrate flux, ΔP is the applied pressure, μ is the filtrate viscosity, R_m is the membrane resistance, α is the specific cake resistance, and m is the cake mass per unit membrane area.

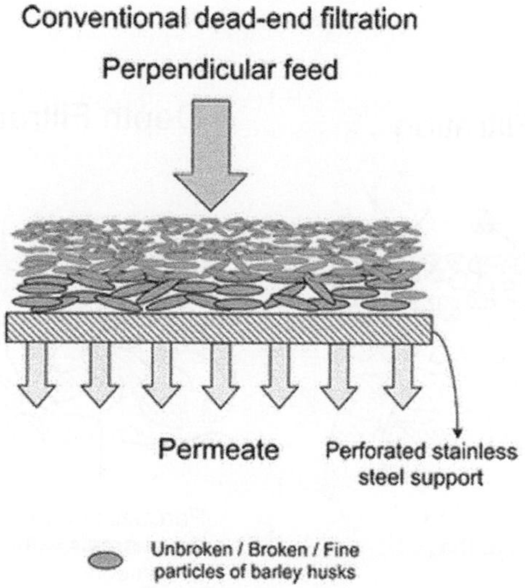

FIGURE 3.4 Mechanism of dead-end filtration. (From Ambrosi, A. et al., *Food Bioprocess Tech.*, 7, 921–936, 2014.)

In dead-end filtration, m is expressed through relation to the filtrate volume, V:

$$m = \frac{cV}{A} \tag{3.2}$$

where A is the membrane area and c is the mass of solid particles per unit volume of filtrate.

The specific cake resistance is measured through the following expression for constant pressure filtration (Ní Mhurchú and Foley, 2006):

$$\frac{t}{V} = \frac{\mu R_m}{A\Delta P} + \frac{\alpha\mu c}{2A^2\Delta P}V \tag{3.3}$$

The specific cake resistance is a measure of "filterability" in filtration operations and also correlates to the efficiency of the process.

The types of filters commonly used in dead-end filtration are depth filters and screen filters. As shown in Figure 3.5, the depth filters do not have a precise pore size or structure; thus, they are not absolute. This means that larger particles will permeate through the filter. Components that are larger than the apertures of the filter will be trapped on the surface of the filter. In the case of smaller particles, random entrapment and adsorption of matter occurs within the structure of the media. Depth filters may also have an electric charge that aids in the entrapment of the smaller particles. The thicker construction and higher porosity of depth filters have led to some advantages, such as higher flow rate and dirt-loading capacity, compared to screens and membrane filters. They are also cheaper than screen and membrane filters. Depth filters are manufactured from fibrous materials, and woven and nonwoven polymeric material or inorganic materials (Krupp et al., 2017). For screen filters, particles are retained directly on the surface of the screen, as shown in Figure 3.5. Unlike depth filters, the pore size of

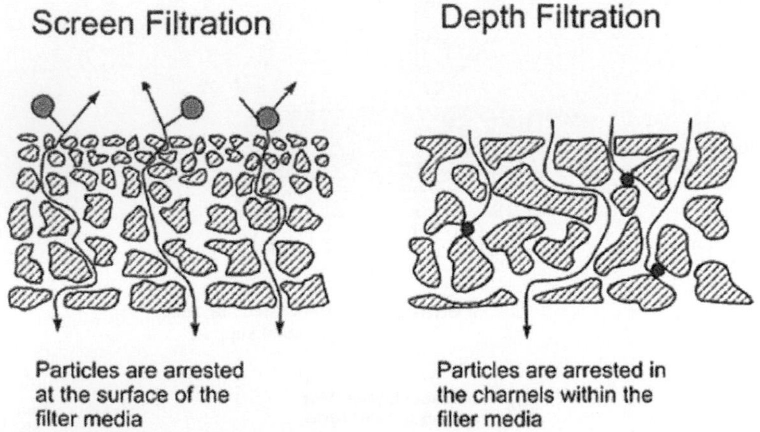
Screen Filtration
Depth Filtration
Particles are arrested at the surface of the filter media
Particles are arrested in the charnels within the filter media

FIGURE 3.5 Mechanisms of screen and depth filtration. (From Shantanu, S., https://www.steviashantanu.com/stevia-membrane-based-extraction.)

the filter is defined precisely. Therefore, only the particles with diameters less than the pore size of the filter can permeate through the filter. Screen filters are used when low nonspecific binding or low adsorption or absorption of the filtrate is needed.

3.3.2 Cross-Flow Filtration

Improvements made to conventional filtration about 40 years ago means that filtration systems are better able to maintain stable filtration rates across the filter medium, known as cross-flow filtration (Bhave, 2014). A majority (more than 98%) of the cross-flow techniques use polymer-based membranes due to the durability and chemical resistant properties in the filtration applications. This has made cross-flow technology relatively cost effective compared to the methods more than forty years ago where polymer chemistry was still not developed.

This technique is known as cross-flow because the direction of flowing streams is parallel to the filter surface; this is entirely opposite to the dead-end filtration technique, where the flow direction is perpendicular to the filter surface (Tien and Ramarao, 2017). The theoretical principles of cross-flow come from addressing the migration of suspended solids in a flowing stream toward a filter surface and potential back-diffusion into the bulk stream, which is derived from Fick's law of diffusion. The velocity of the fluid flowing parallel to the surface of the filter medium becomes a control of the suspended solids' concentration at the surfaces of the filter medium, as shown in Figure 3.6. Designing a favorable cross-flow system depends on factors like selecting an appropriate membrane geometry that could give consistent results; compatibility of the membrane with chemicals used during cleaning; and, most important, economical installation and operation (Lorente et al., 2017).

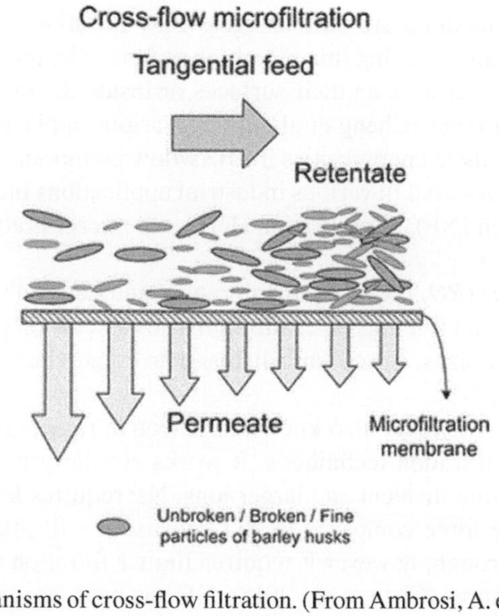

FIGURE 3.6 Mechanisms of cross-flow filtration. (From Ambrosi, A. et al., *Food Bioprocess Tech.*, 7, 921–936, 2014.)

The filtration flux, *J*, as a function of transmembrane pressure drop is shown as the following expression (Shuler et al., 2017):

$$J = \frac{\Delta P_M}{R_G + R_M} \tag{3.4}$$

where R_G is the gel resistance (the value of R_G varies with solute concentration and cross-flow velocity across the membrane) and R_M is the membrane resistance, a value that is constant.

The average transmembrane pressure drop is (Shuler et al., 2017):

$$\Delta P_M = P_i - \frac{1}{2}\Delta P \tag{3.5}$$

Expression of pressure drop for laminar flow follows Hagen-Poiseuille equation (Shuler et al., 2017):

$$\Delta P = \frac{C_1 \mu L V}{d^2} = \frac{C_2 \mu L Q}{d^4} \tag{3.6}$$

The expression of pressure drop for turbulent flow is:

$$\Delta P = \frac{C_3 \mu L V^2}{d} = \frac{C_4 f L Q^2}{d^5} \tag{3.7}$$

where *f* is the Fanning friction factor correlates to function of *Re*.

In cross-flow filtration techniques, turbulent flow is preferred compared to laminar (Shuler et al., 2017).

Membranes filter medium are used in cross-flow filtration. Polymeric materials are usually used in manufacturing thin membrane filters. The main function of membrane filters is to trap particles on their surfaces or inside the membranes according to the sizes of the particles (Cheng et al., 2018). Various applications are addressed with tailored range of different porosities in cross-flow membrane filtration. Common examples of membranes used in various industrial applications include reverse osmosis (RO), nanofiltration (NF), ultrafiltration (UF), and microfiltration (MF):

1. *Reverse osmosis (RO)*: RO is the membrane manufactured with the smallest pores. RO does not filter particles through a size exclusion process based on its smallest pore sizes; it uses ionic diffusion to create the separation instead (Alonso et al., 2018).
2. *Nanofiltration (NF)*: NF, also known as selective rejection, is the latest of the cross-flow filtration techniques. It works similarly to RO but focuses more on removing divalent and larger ions. NF requires less osmotic pressure as driving force compared to RO because it still allows monovalent ions to pass through, however it requires higher filtration rates with lower operating pressures (Chen et al., 2018). NF has been used to replace RO is many applications today.

3. *Ultrafiltration (UF)*: UF filters particles based on molecular weight. Its rating is measured in molecular weight cut-off (MWCO) instead of pore sizes (Guo et al., 2014). UF filtration is often employed in the concentration, removal steps and diafiltration within downstream processes of pharmaceutical industries (Wei et al., 2016).
4. *Microfiltration (MF)*: MF can be employed with dead-end or cross-flow filtration. It is known as a hybrid filtration system. However, some applications will not be fully optimized using MF operating in the dead-end mode, especially in processes that involve large amounts of insoluble materials. Thus, most MF processes were performed in the direction of cross-flow. MF is commonly used with various separation processes like ultrafiltration and reverse osmosis (Morales et al., 2018). It serves as post-treatment and pretreatment to granular media filtration and separation processes like ultrafiltration, respectively (Brandt et al., 2017).

3.3.3 Industrial Applications

Dead-end filtration techniques are commonly used in concentrating compounds. Dead-end filtration techniques effectively remove particles with low concentration and low pressure drop across the filter medium due to the packing tendency. Dead-end filter media are also easier and cheaper to build and easier to use, making the entire technique cheaper than cross-flow filtration (Winans et al., 2016). Filtration rate drops over time due to high concentrations of particles on process streams such as cells and precipitates compacting on the filter surface in dead-end mode. Hence, dead-end filtration requires removal of accumulated particles clogging the filter after a period of time, making this a batch process. Dead-end filtration is in common use, including home water filtration systems; vacuum cleaners; and sterile filtration of wine, beer, and water in typical industrial applications.

Cross-flow membrane filtration is very common in many industries all over the world. The principles used in this technology are reverse osmosis, nanofiltration, ultrafiltration, and microfiltration. Each of these principles has various applications in different targeted industries, and some may overlap, where more than one principle could be applied in the same industry to serve different purposes. A common example is membrane filtration in purifying water. In fact, all four types of filtration methods are suitable to purify water. Hence, an individual industry has to decide which types of filters are preferred according to their own processes to address the problem of getting clean water. More applications of each principle are explained below.

1. *Reverse osmosis*: The major application for reverse osmosis (RO) is the production of fresh water for industries. Even household drinking water purification systems around the world uses RO systems to improve water quality for consumption and cooking. Other freshwater applications include water and wastewater purification, where industry uses RO systems to obtain pure water by removing minerals from boiler water at power plants. The purpose is to avoid mineral deposition, which leads to underperformance

of the boiler and directly affect its efficiency (Suárez et al., 2014). RO is commonly used in food industries, especially in dairy and wine industries, to concentrate food liquids because it is more economical compared to conventional heat-treatment processes.

2. *Nanofiltration*: Nanofiltration (NF) was used only in molecular separation in the early days. Now its application has extended to a wider range of industries such as juice and milk production, pharmaceuticals, fine chemicals, and flavor/fragrance industries (Nath et al., 2018). NF does the job of product polishing and recovery of homogeneous catalysts in the industry of bulk chemistry production. Some common examples of applications using NF is the removal of tar components of the feed streams and purification of gas condensates in petroleum industries. NF is very helpful in extracting amino acids and lipids from cell cultures in the field of medicine (Ji et al., 2017).
3. *Ultrafiltration*: Ultrafiltration (UF) is commonly used in purifying water, where it removes macromolecules and particulates to produce potable water. UF is extensively applied in dairy production, particularly in obtaining whey protein concentrate. The once-existing method of obtaining whey protein concentrate was through steam heating followed by drum or spray drying. However, this method gives inconsistency in product composition and requires high capital and operating cost due to the excessive heat used to dry the product, and drying at high temperature also denatures some of the protein content (Tamime, 2012). Thus, most of the dairy industries has switched to the use of UF in concentrating whey protein.
4. *Microfiltration*: Microfiltration (MF) is a crucial method in cold sterilization for pharmaceutical and beverage industries. It is crucial because the existing method employs heat treatments in sterilizing the products, causing loss of flavor in beverages and loss of effectiveness in drugs. Sterilization using MF membranes avoids the use of heat and overcomes the problems faced by traditional methods. Other common applications of MF are purification of cell broths in separating macromolecules from larger molecules as well as production of adhesives and paints (Baker, 2012; Starbard, 2008).

3.3.4 Advantages and Disadvantages

3.3.4.1 Dead-End Filtration versus Cross-Flow Filtration

Both dead-end and cross-flow filtration technologies serve the purpose of purifying bioprocess solutions by removing contaminants. Each technique has unique advantages and drawbacks when it comes to the separation process. In general, dead-end filtration is used when clarification is needed in relatively low solid streams, final polishing is to be done in achieving certain sterility, and protection and enhancement of downstream operations are required. Cross-flow filtrations are employed for higher solid and more viscous feed streams where purification and concentration of targeted species are required. Table 3.4 compares dead-end and cross-flow filtration modes.

TABLE 3.4
Comparison of Dead-End and Cross-Flow Filtration

Method	Advantages	Disadvantages
Dead-end filtration	• Almost 100% of collection rate • Possible on miniaturized scale • Low cost • Does not require backwash/ chemical cleaning	• Filters need to be replaced often due to the formation of filter cakes • Fail to perform separation when large amount of insoluble materials present
Cross-flow filtration	• Requires less filter maintenance • Can be used even when large amount of insoluble materials present • Can be used on viscous liquids • Can be reused by backwashing/ chemical cleaning	• Lower collection rate • High cost • Pieces of equipment are usually larger and more complicated in operation

3.3.4.2 Filtration versus Other Separation Methods

Recent developments have shown techniques with better separation efficiency compared to filtration, such as chromatography and centrifugation separation. However, the filtration techniques, especially cross-flow filtration, remain one of the most useful technologies for both gas and solid-liquid separation. Table 3.5 summarizes the differences between the filtration method and other types of separation techniques.

TABLE 3.5
Comparison of Filtration with Other Separation Techniques

Methods	Advantages	Disadvantages
Filtration (crossflow filtration)	• Highly effective in fractionating particle species according to size • Energy requirements are low • Processes are relatively simple to scale up	• Processes are prone to membrane fouling effects (expensive cleaning) • Equipment cost could be high
Chromatography	• Capable of separating materials according to size and chemical properties • Can be used to separate delicate or heat-labile compounds	• Process scale-up is a problem (low throughput) • Prefiltration of feed material is usually required
Centrifugation	• Can be selected for different applications • Has more process flexibility and higher levels of performance	• High energy consumption • High initial capital cost
Ion exchange	• Removes dissolved inorganics effectively • Regenerable • Relatively inexpensive initial capital investment	• Does not effectively remove particles, pyrogens, or bacteria • High operating costs over long term

3.4 ADVANCED DEVELOPMENT IN FILTRATION

Most modern industrial facilities that use filtration typically do not rely on just filtration alone. Various technologies are used in conjunction with filtration to enhance the efficiency and performance of a separation process. For example, recent developments in nanotechnology have piqued interest in applying nanomaterials for wastewater treatment. Filtration membranes can be modified to include nanoparticles that may improve the characteristics of the membrane. A study has demonstrated that ultrafiltration membranes modified with alumina nanoparticles can increase membrane strength and fouling resistance (Maximous et al., 2010). Carbon nanotubes have antimicrobial attributes that can inhibit biological growth on the membrane, thus giving it a longer lifespan (Mauter et al., 2011). Nanoparticles have numerous benefits in terms of filtration performance, but the potential effects of using nanomaterials on the environment and human health has not been studied properly. Several studies have shown that particles at the nanoscale may have different characteristics than their macro-sized counterparts, and nanoparticles may cause adverse effects to human health and the environment compared to bulk particles (Ray et al., 2009; Albanese et al., 2012). In addition to the high costs of nanoparticles, a better understanding of the risks involved is required before the integration of nanotechnology with water treatment technology can be widely implemented (Marambio-Jones and Hoek, 2010; Qu et al., 2013).

Today, the integration of advanced water treatment technology used in conjunction with conventional filtration techniques is common for enhancing the effectiveness of water purification. New waterborne pathogens continue to emerge, making disinfection of water increasingly challenging because new techniques need to be developed to treat and eliminate these pathogens. A study has mentioned that certain waterborne pathogens such as *C. parvum* have great resistance to chlorine (Shannon et al., 2008). It was observed that ultraviolet (UV) light is effective in dealing with such pathogens because UV can photochemically inactivate these microbes. An increased number of research and development studies into the use of membrane bioreactors (MBRs) in water purification shows that MBR combines the use of microfiltration or ultrafiltration membranes and the suspension of biomass to clarify wastewater. One of the most promising applications of MBR is the multistage treatment process that uses MBR as a pretreatment stage prior to reverse osmosis, followed by UV disinfection. This effectively produces potable water free of suspended solids and pathogens. The process of producing potable water through MBR is shown in Figure 3.7.

The main limitation of MBR is that the membranes are highly susceptible to fouling, which causes a permanent loss in flux and cannot be remedied through cleaning (Kimura et al., 2005). Therefore, further development of cheaper and nonfouling membranes with a high flux is needed for extensive use in the industry. The use

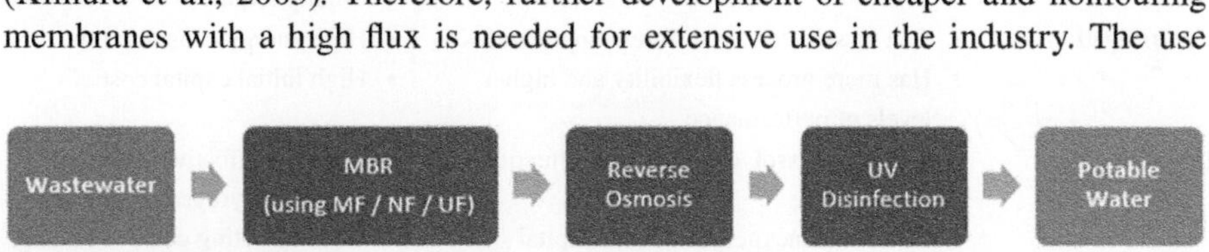

FIGURE 3.7 Membrane bioreactor (MBR) water treatment system.

of photocatalysts such as titanium dioxide (TiO_2) combined with filtration in water treatment has shown a lot of promise as researchers have been carrying out extensive studies on photocatalytic water treatment in recent years (Lazar et al., 2012; Athanasekou et al., 2015). This method is essentially the same as the UV disinfection process except that photocatalysts are added in a suspension or are immobilized on an inert support. The photocatalysts are activated in the presence of UV light and then react with the contaminants to break them down into carbon dioxide and water (Chong et al., 2010).

As previously mentioned in this chapter, there are two conventional modes of filtration: dead-end filtration and cross-flow filtration. A newly developed mode of filtration known as dynamic crossflow filtration (DCF) uses dynamic or rotating membranes in cross-flow filtration. Figure 3.8 shows the mechanism of dynamic cross-flow filtration compared to dead-end filtration and traditional cross-flow filtration.

Conventional cross-flow filtration relies on a feed flow that is tangential to the membrane to prevent the accumulation of solids on the membrane, which may lead to fouling. DCF has the advantage of increasing permeate flux due to the shearing effect at high rotation speeds in addition to reducing the cake formation and increasing resistance to fouling. Other filtration membranes such as ultrafiltration, nanofiltration, and reverse osmosis may be used along with DCF if desired. DCF is especially useful for processes involving highly viscous liquids as the retentate can be concentrated until it exits through the outlet (Hoek, 2013).

A study of the DCF units developed by Novoflow has shown that they are very effective compared to conventional cross-flow filtration (Liebermann, 2010). The design of the Novoflow cross-flow filtration system consists of a hollow chamber that contains the solid-liquid mixture with a shaft in the center. The shaft holds the filter disks, which will be rotating inside like an impeller of a pump. Figure 3.9 shows a lab-scale DCF unit by Novoflow. Liebermann states that there is significant cost

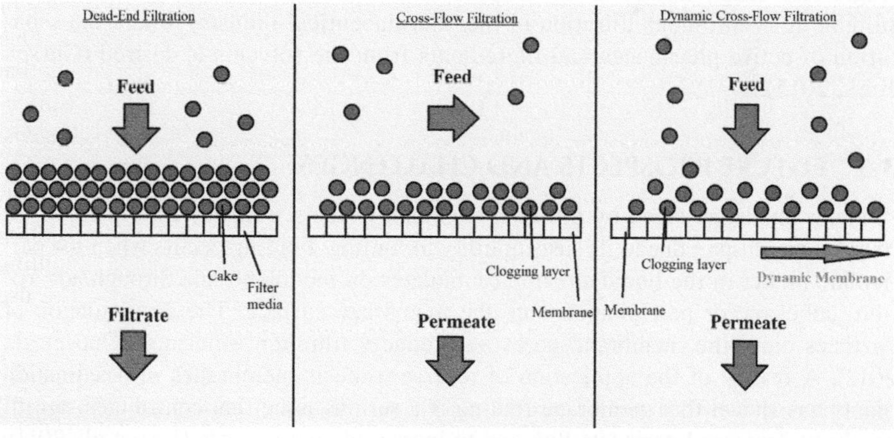

FIGURE 3.8 Mechanism of dynamic cross-flow filtration compared to conventional filtration modes.

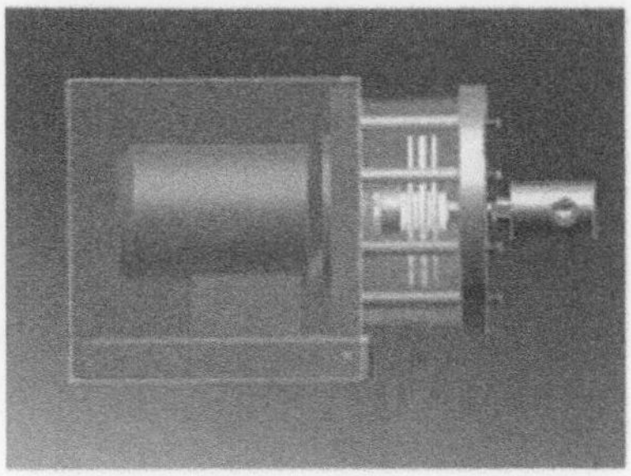

FIGURE 3.9 Lab-scale Novoflow DCF. (From Liebermann, F., *Desalination*, 250, 1087–1090, 2010.)

savings due to the rotating action of the rotating filter disks, which creates a centrifugal force and thus negates the need of a feed inlet pump. This allows the system to be more efficient at handling denser and more viscous solid-liquid mixtures because the rotation energy required is lower than the energy required for the operation of feed pumps. Another key advantage of DCF is that it is possible to use various types of filter disks for the system, depending on the application required. For example, ceramic filter disks with pore sizes as small as 0.1 μm were used to carry out microfiltration or ultrafiltration. Although ceramic filter disks have excellent chemical and thermal stability, they are not used for intensive applications since they are brittle and costly to manufacture.

DCF systems such as the one designed by Novoflow have been implemented in wastewater processing where they can be retrofitted into existing plant setups to conduct large-scale water treatment. Moreover, such systems are beneficial in the semiconductor industry because they are very effective in the recovery of valuable solids. The rotating filters can retain high concentration of solids in a single step (Liebermann, 2010). A study has demonstrated that DCF is highly suitable for continuous filtration in the pharmaceutical industry where the separation of active pharmaceutical ingredients from the solvents is desired (Gursch et al., 2015).

3.5 FUTURE PROSPECTS AND CHALLENGES

A major challenge faced by filtration technology stems from the filter medium or membrane lifespan due to its susceptibility to fouling. Fouling occurs when the particulate matter in the liquid stream accumulates on the membrane through adsorption, adhesion, or precipitation onto the membrane surface. The accumulation of particles clogs the membrane pores and reduces filtration efficiency (Guo et al., 2012). A review of the application of reverse osmosis membranes in desalination plants has shown that membrane fouling is a serious issue that contributes significantly to decreased permeate flux and to increased energy costs (Lee et al., 2011). The development of improved RO membranes with higher efficiencies has made the desalination process more economical over the years. However, research efforts are still being made to find an ideal RO membrane that possesses high permeability and high ion rejection with a strong structural integrity. An ideal RO membrane with those characteristics will greatly reduce the capital and operational costs of the reverse osmosis technology.

The membrane of any filtration process that involves biological effluents or liquids that contain microbial matter is also subject to biofouling. The growth and deposition of microbial matter on the membrane can be partially inhibited through the addition of polymers and nanoparticles on the membrane surface to modify its surface properties (Mansouri et al., 2010). However, this only serves as a temporary solution as it is not mechanically or chemically stable. Further development in this area may potentially bring countless advantages such as lowering maintenance costs and increasing the life cycle of the membrane.

New types of filtration media are being develop continuously to further improve membrane properties and filtration performance. Many scientists have focused on nanotechnology to develop with filter membranes with nanomaterials. However, research on filtration and membrane technology combined with nanotechnology is still in its early stages. Several challenges including its high material costs and the difficulty in scaling up, which will inevitably prolong the time before such membranes can be used commercially worldwide (Lee et al., 2011). More recently, researchers have produced carbon nanotube (CNT) membranes to investigate their performance in water desalination. Their work has demonstrated that CNT membranes possess many advantages, such as excellent desalination capacity, selectivity of solutes, and antifouling properties (Das et al., 2014).

The use of ceramic membranes has often been limited to small-scale separation processes because of the high cost of production (Pendergast and Hoek, 2011). Several researchers have previously demonstrated the successful preparation of economical ceramic membranes that can be used in microfiltration and ultrafiltration (Nandi et al., 2008). If this process can be scaled up to a large-scale production of low-cost ceramic membranes, then the use of such ceramic membranes may potentially extend in the future to processes with large-scale separation processes such as water treatment plants.

One of the challenges in air filtration is balancing particle penetration with the resistance to air flow because a finer membrane will inherently have a larger pressure drop across the membrane due to the obstruction to air flow. The article titled "Filtration and Separation" mentioned that high-performance air filters like the ultra-high molecular weight polyethylene (UPE) membrane filter has been developed for high filtration efficiency while possessing low resistance to air flow (Galka and Saxena, 2009). UPE membranes are inherently strong because they are designed to be thin with high particle loading capacity. These membranes are the most useful where high efficiency and low air resistance is desired, such as respiratory devices for emergency first responders and in industrial cleanrooms. A novel filtration membrane was developed using electrospun polymer nanofibers, which are optimized for filtering aerosol particles; they serve as proof that there is immense potential in nanomaterial air filtration membranes (Yun et al., 2010). As more companies continue to design and develop better filter membranes, air filtration technology will become more and more promising.

3.6 CONCLUSION

Filtration is one of the best methods to conduct gas and solid-fluid separation. Conventional filtration methods such as dead-end filtration and cross-flow filtration remain widely used today in many industries. New and improved filter media are being developed continuously for enhanced efficiency and performance. Novel and innovative approaches to filtration such as dynamic cross-flow filtration serves as proof that filtration has a bright and promising future. The main principle of utilizing a filter medium to carry out the separation of solids and fluids has remained the same for thousands of years and will likely stay that way for the time being.

ABBREVIATIONS

ASHRAE	American Society of Heating Refrigerating and Air-Conditioning engineers
ASME	American Society of Mechanical Engineers
BS	British Standard
CEN	European Committee for Standardization
CNTs	carbon nanotubes
DCF	dynamic crossflow filtration
DIN	Deutsches Institute for Normung (German Institute for Standardization
EN	European Standards
ESPs	electrostatic precipitators
HEPA	high-efficiency particulate air
HPTFF	high-performance tangential flow filtration
MBRs	membrane bioreactors
MF	microfiltration
MWCO	molecular weight cut-off
NF	nanofiltration
RO	reverse osmosis
TFF	tangential flow filtration
UF	ultrafiltration
ULPA	ultra-low penetration air
UPE	ultra-high molecular weight polyethylene
VOCs	Volatile organic compounds

REFERENCES

Albanese, A., P. S. Tang, and W. C. W. Chan. 2012. "The effect of nanoparticle size, shape, and surface chemistry on biological systems." *Annual Review of Biomedical Engineering* 14 (1):1–16. doi:10.1146/annurev-bioeng-071811-150124.

Alkemade, U. G., and B. Schumann. 2006. "Engines and exhaust after treatment systems for future automotive applications." *Solid State Ionics* 177 (26):2291–2296. doi:10.1016/j.ssi.2006.05.051.

Alonso, J. J. S., N. El Kori, N. Melián-Martel, and B. Del Río-Gamero. 2018. "Removal of ciprofloxacin from seawater by reverse osmosis." *Journal of Environmental Management* 217:337–345. doi:10.1016/j.jenvman.2018.03.108.

Ambrosi, A., N. S. M. Cardozo, and I. C. Tessaro. 2014. "Membrane separation processes for the beer industry: A review and state of the art." *Food and Bioprocess Technology* 7 (4):921–936. doi:10.1007/s11947-014-1275-0.

Athanasekou, C. P., N. G. Moustakas, S. Morales-Torres, L. M. Pastrana-Martínez, J. L. Figueiredo, J. L. Faria, A. M. T. Silva, J. M. Dona-Rodriguez, G. E. Romanos, and P. Falaras. 2015. "Ceramic photocatalytic membranes for water filtration under UV and visible light." *Applied Catalysis B: Environmental* 178:12–19. doi:10.1016/j.apcatb.2014.11.021.

Baker, R. W. 2012. "Microfiltration." In *Membrane Technology and Applications* (3rd ed.), edited by R. W. Baker, pp. 303–324. West Sussex, UK: John Wiley and Sons Ltd.

Baldasso, C., T. C. Barros, and I. C. Tessaro. 2011. "Concentration and purification of whey proteins by ultrafiltration." *Desalination* 278 (1):381–386. doi:10.1016/j.desal.2011.05.055.

Bekö, G., G. Clausen, and C. J. Weschler. 2008. "Is the use of particle air filtration justified? Costs and benefits of filtration with regard to health effects, building cleaning and occupant productivity." *Building and Environment* 43 (10):1647–1657. doi:10.1016/j.buildenv.2007.10.006.

Bennett, A. 2015. "Water & wastewater: Advances in filtration systems for wastewater treatment." *Filtration + Separation* 52 (5):28–33. doi:10.1016/S0015-1882(15)30222-6.

Bennett, A. 2016. "Developments in air & gas filtration technology." *Filtration + Separation* 53 (5):30–35. doi:10.1016/S0015-1882(16)30210-5.

Bhave, R. R. 2014. "Chapter 9—Cross-flow filtration A2—Vogel, Henry C." In *Fermentation and Biochemical Engineering Handbook* (3rd ed.), edited by C. M. Todaro, pp. 149–180. Boston, MA: William Andrew Publishing.

Boltic, Z., N. Ruzic, M. Jovanovic, M. Savic, J. Jovanovic, and S. Petrovic. 2013. "Cleaner production aspects of tablet coating process in pharmaceutical industry: Problem of VOCs emission." *Journal of Cleaner Production* 44:123–132. doi:10.1016/j.jclepro.2013.01.004.

Brandt, M. J., K. M. Johnson, A. J. Elphinston, and D. D. Ratnayaka. 2017. "Chapter 9—Water filtration." In *Twort's Water Supply* (7th ed.), pp. 367–406. Boston, MA: Butterworth-Heinemann.

Bull, K. 2008. "Cabin air filtration: Helping to protect occupants from infectious diseases." *Travel Medicine and Infectious Disease* 6 (3):142–144. doi:10.1016/j.tmaid.2007.08.004.

Chen, X., Y. Zhang, J. Tang, M. Qiu, K. Fu, and Y. Fan. 2018. "Novel pore size tuning method for the fabrication of ceramic multi-channel nanofiltration membrane." *Journal of Membrane Science* 552:77–85. doi:10.1016/j.memsci.2018.01.056.

Cheng, X. Q., Z. X. Wang, X. Jiang, T. Li, C. H. Lau, Z. Guo, J. Ma, and L. Shao. 2018. "Towards sustainable ultrafast molecular-separation membranes: From conventional polymers to emerging materials." *Progress in Materials Science* 92:258–283. doi:10.1016/j.pmatsci.2017.10.006.

Chew, A. W. Z., and A. W. K. Law. 2018. "DRFM hybrid model to optimize energy performance of pre-treatment depth filters in desalination facilities." *Applied Energy* 220:576–597. doi:10.1016/j.apenergy.2018.03.028.

Chong, M. N., B. Jin, C. W. K. Chow, and C. Saint. 2010. "Recent developments in photocatalytic water treatment technology: A review." *Water Research* 44 (10):2997–3027. doi:10.1016/j.watres.2010.02.039.

Das, D., and A. Waychal. 2016. "On the triboelectrically charged nonwoven electrets for air filtration." *Journal of Electrostatics* 83:73–77. doi:10.1016/j.elstat.2016.08.004.

Das, R., M. E. Ali, S. B. A. Hamid, S. Ramakrishna, and Z. Z. Chowdhury. 2014. "Carbon nanotube membranes for water purification: A bright future in water desalination." *Desalination* 336:97–109. doi:10.1016/j.desal.2013.12.026.

Dixit, M., and U. Braeutigam. 2007. "Biopharmaceutical industry: The importance of pre-filtration." *Filtration & Separation* 44 (6):24–26. doi:10.1016/S0015-1882(07)70181-7.

dos Santos Mathias, T. R., P. P. M. de Mello, and E. F. C. Sérvulo. 2014. "Solid wastes in brewing process: A review." *Journal of Brewing and Distilling* 5 (1). doi:10.5897/JBD2014.0043.

Ellouze, E., N. Tahri, and R. B. Amar. 2012. "Enhancement of textile wastewater treatment process using Nanofiltration." *Desalination* 286:16–23. doi:10.1016/j.desal.2011.09.025.

Gadipelly, C., A. Pérez-González, G. D. Yadav, I. Ortiz, R. Ibáñez, V. K. Rathod, and K. V. Marathe. 2014. "Pharmaceutical industry wastewater: Review of the technologies for water treatment and reuse." *Industrial & Engineering Chemistry Research* 53 (29):11571–11592. doi:10.1021/ie501210j.

Galka, N., and A. Saxena. 2009. "High efficiency air filtration: The growing impact of membranes." *Filtration & Separation* 46 (4):22–25. doi:10.1016/S0015-1882(09)70157-0.

Gantz, C. 2012. *The Vacuum Cleaner: A History.* McFarland, CA: Incorporated, Publishers.

Ge, X., D. S. C. Yang, K. Trabbic-Carlson, B. Kim, A. Chilkoti, and C. D. M. Filipe. 2005. "Self-cleavable stimulus responsive tags for protein purification without chromatography." *Journal of the American Chemical Society* 127 (32):11228–11229. doi: 10.1021/ja0531125.

Gormley, T., T. A. Markel, H. Jones, D. Greeley, J. Ostojic, J. H. Clarke, M. Abkowitz, and J. Wagner. 2017. "Cost-benefit analysis of different air change rates in an operating room environment." *American Journal of Infection Control* 45 (12):1318–1323. doi:10.1016/j.ajic.2017.07.024.

Grzenia, D. L., J. O. Carlson, and S. R. Wickramasinghe. 2008. "Tangential flow filtration for virus purification." *Journal of Membrane Science* 321 (2):373–380. doi:10.1016/j.memsci.2008.05.020.

Guo, J., H. Liu, J. Liu, and L. Wang. 2014. "Ultrafiltration performance of EfOM and NOM under different MWCO membranes: Comparison with fluorescence spectroscopy and gel filtration chromatography." *Desalination* 344:129–136. doi:10.1016/j.desal.2014.03.006.

Guo, W., H. H. Ngo, and J. Li. 2012. "A mini-review on membrane fouling." *Bioresource Technology* 122:27–34. doi:10.1016/j.biortech.2012.04.089.

Gursch, J., R. Hohl, G. Toschkoff, D. Dujmovic, J. Brozio, M. Krumme, N. Rasenack, and J. Khinast. 2015. "Continuous processing of active pharmaceutical ingredients suspensions via dynamic cross-flow filtration." *Journal of Pharmaceutical Sciences* 104 (10):3481–3489. doi:10.1002/jps.24562.

Hjorth, M., K. V. Christensen, M. L. Christensen, and S. G. Sommer. 2011. "Solid–Liquid separation of animal slurry in theory and practice." In *Sustainable Agriculture Volume 2*, edited by E. Lichtfouse, M. Hamelin, M. Navarrete, and P. Debaeke, 953–986. Dordrecht, the Netherlands: Springer Netherlands.

Hoek, E. M., V. V. Tarabara, and M. Y. Jaffrin. 2013. "Dynamic crossflow filtration." In *Encyclopedia of Membrane Science and Technology Volume 2*, edited by E. M. V. Hoek and V. V. Tarabara, pp. 1190–1218. Hoboken, NJ: Wiley.

Jenkins, M. W., S. K. Tiwari, and J. Darby. 2011. "Bacterial, viral and turbidity removal by intermittent slow sand filtration for household use in developing countries: Experimental investigation and modeling." *Water Research* 45 (18):6227–6239. doi:10.1016/j.watres.2011.09.022.

Ji, Y., W. Qian, Y. Yu, Q. An, L. Liu, Y. Zhou, and C. Gao. 2017. "Recent developments in nanofiltration membranes based on nanomaterials." *Chinese Journal of Chemical Engineering* 25 (11):1639–1652. doi:10.1016/j.cjche.2017.04.014.

Kimura, K., N. Yamato, H. Yamamura, and Y. Watanabe. 2005. "Membrane fouling in pilot-scale membrane bioreactors (MBRs) treating municipal wastewater." *Environmental Science & Technology* 39 (16):6293–6299. doi:10.1021/es0502425.

Krupp, A. U., C. P. Please, A. Kumar, and I. M. Griffiths. 2017. "Scaling-up of multi-capsule depth filtration systems by modeling flow and pressure distribution." *Separation and Purification Technology* 172:350–356. doi:10.1016/j.seppur.2016.07.028.

Lazar, M. A., S. Varghese, and S. S Nair. 2012. "Photocatalytic water treatment by titanium dioxide: Recent updates." *Catalysts* 2 (4):572–601.

Lebreton, B., A. Brown, and R. van Reis. 2008. "Application of high-performance tangential flow filtration (HPTFF) to the purification of a human pharmaceutical antibody fragment expressed in *Escherichia coli*." *Biotechnology and Bioengineering* 100 (5): 964–974. doi:10.1002/bit.21842.

Lee, K. P., T. C. Arnot, and D. Mattia. 2011. "A review of reverse osmosis membrane materials for desalination—Development to date and future potential." *Journal of Membrane Science* 370 (1):1–22. doi:10.1016/j.memsci.2010.12.036.

Liebermann, F. 2010. "Dynamic cross flow filtration with Novoflow's single shaft disk filters." *Desalination* 250 (3):1087–1090. doi:10.1016/j.desal.2009.09.114.

Lipnizki, F. 2010. "Cross-flow membrane applications in the food industry." In *Membrane Technology Volume 3*, edited by K-V. Peinemann, S. P. Nunes, and L. Giorno, pp. 1–24. Weinheim, Germany: Wiley-VCH Verlag GmbH & Co. KGaA.

Liu, G., M. Xiao, X. Zhang, C. Gal, X. Chen, L. Liu, S. Pan, J. Wu, L. Tang, and D. Clements-Croome. 2017. "A review of air filtration technologies for sustainable and healthy building ventilation." *Sustainable Cities and Society* 32:375–396. doi:10.1016/j.scs.2017.04.011.

Liu, Z., Z. Ji, X. Wu, H. Ma, F. Zhao, and Y. Hao. 2016. "Experimental investigation on liquid distribution of filter cartridge during gas-liquid filtration." *Separation and Purification Technology* 170:146–154. doi:10.1016/j.seppur.2016.06.032.

Lorente, E., M. Hapońska, E. Clavero, C. Torras, and J. Salvadó. 2017. "Microalgae fractionation using steam explosion, dynamic and tangential cross-flow membrane filtration." *Bioresource Technology* 237:3–10. doi:10.1016/j.biortech.2017.03.129.

Maddineni, A. K., D. Das, and R. M. Damodaran. 2017. "Inhibition of particle bounce and re-entrainment using oil-treated filter media for automotive engine intake air filtration." *Powder Technology* 322:369–377. doi:10.1016/j.powtec.2017.09.025.

Mah, S. K., C. C. H. Chang, T. Y. Wu, and S. P. Chai. 2014. "The study of reverse osmosis on glycerin solution filtration: Dead-end and crossflow filtrations, transport mechanism, rejection and permeability investigations." *Desalination* 352:66–81. doi:10.1016/j.desal.2014.08.008.

Mansouri, J., S. Harrisson, and V. Chen. 2010. "Strategies for controlling biofouling in membrane filtration systems: Challenges and opportunities." *Journal of Materials Chemistry* 20 (22):4567–4586. doi:10.1039/B926440J.

Marambio-Jones, C., and E. M. V. Hoek. 2010. "A review of the antibacterial effects of silver nanomaterials and potential implications for human health and the environment." *Journal of Nanoparticle Research* 12 (5):1531–1551. doi:10.1007/s11051-010-9900-y.

Mauter, M. S., Y. Wang, K. C. Okemgbo, C. O. Osuji, E. P. Giannelis, and M. Elimelech. 2011. "Antifouling ultrafiltration membranes via post-fabrication grafting of biocidal nanomaterials." *ACS Applied Materials & Interfaces* 3 (8):2861–2868. doi:10.1021/am200522v.

Maximous, N., G. Nakhla, K. Wong, and W. Wan. 2010. "Optimization of Al2O3/PES membranes for wastewater filtration." *Separation and Purification Technology* 73 (2): 294–301. doi:10.1016/j.seppur.2010.04.016.

Morales, D., F. R. Smiderle, A. J. Piris, C. Soler-Rivas, and M. Prodanov. 2018. "Production of a β-d-glucan rich extract from Shiitake mushrooms (Lentinula edodes) by an extraction/microfiltration/reverse osmosis (nanofiltration) process." *Innovative Food Science & Emerging Technologies*. doi:10.1016/j.ifset.2018.04.003.

Nandi, B. K., R. Uppaluri, and M. K. Purkait. 2008. "Preparation and characterization of low cost ceramic membranes for micro-filtration applications." *Applied Clay Science* 42 (1):102–110. doi:10.1016/j.clay.2007.12.001.

Nath, K., H. K. Dave, and T. M. Patel. 2018. "Revisiting the recent applications of nanofiltration in food processing industries: Progress and prognosis." *Trends in Food Science & Technology* 73:12–24. doi:10.1016/j.tifs.2018.01.001.

Ní Mhurchú, J., and G. Foley. 2006. "Dead-end filtration of yeast suspensions: Correlating specific resistance and flux data using artificial neural networks." *Journal of Membrane Science* 281 (1):325–333. doi:10.1016/j.memsci.2006.03.043.

Palmer Andre, F., G. Sun, and R. H. David. 2008. "Tangential flow filtration of hemoglobin." *Biotechnology Progress* 25 (1):189–199. doi:10.1002/btpr.119.

Parker Hannifin plc, Filter Division Peel Street, Morley Leeds LS27 8EL, UK. 1993. "Cost savings in pharmaceutical and cosmetic prefiltration." *Filtration and Separation* 30 (1):33–34.

Pendergast, M. M., and E. M. V. Hoek. 2011. "A review of water treatment membrane nanotechnologies." *Energy & Environmental Science* 4 (6):1946–1971. doi:10.1039/C0EE00541J.

Pouliot, Y. 2008. "Membrane processes in dairy technology—From a simple idea to worldwide panacea." *International Dairy Journal* 18 (7):735–740. doi:10.1016/j.idairyj.2008.03.005.

Qu, X., P. J. J. Alvarez, and Q. Li. 2013. "Applications of nanotechnology in water and wastewater treatment." *Water Research* 47 (12):3931–3946. doi:10.1016/j.watres.2012.09.058.

Ray, P. C., H. Yu, and P. P. Fu. 2009. "Toxicity and environmental risks of nanomaterials: Challenges and future needs." *Journal of Environmental Science and Health: Part C, Environmental Carcinogenesis & Ecotoxicology Reviews* 27 (1):1–35. doi:10.1080/10590500802708267.

Rebai, M., M. Prat, M. Meireles, P. Schmitz, and R. Baclet. 2010. "Clogging modeling in pleated filters for gas filtration." *Chemical Engineering Research and Design* 88 (4):476–486. doi:10.1016/j.cherd.2009.08.014.

Saxena, A., B. P. Tripathi, M. Kumar, and V. K. Shahi. 2009. "Membrane-based techniques for the separation and purification of proteins: An overview." *Advances in Colloid and Interface Science* 145 (1):1–22. doi:10.1016/j.cis.2008.07.004.

Schneider, J., M. Krottenthaler, W. Back, and H. Weisser. 2005. "Study on the membrane filtration of mash with particular respect to the quality of wort and beer." *Journal of the Institute of Brewing* 111 (4):380–387. doi:10.1002/j.2050-0416.2005.tb00223.x.

Scott, K., and R. Hughes. 2012. *Industrial Membrane Separation Technology*. Dordrecht, the Netherlands: Springer.

Selatile, M. K., S. S. Ray, V. Ojijo, and R. Sadiku. 2018. "Depth filtration of airborne agglomerates using electrospun bio-based polylactide membranes." *Journal of Environmental Chemical Engineering* 6 (1):762–772. doi:10.1016/j.jece.2017.12.070.

Shahtalebi, A. 2011. "Application of nanofiltration membrane in the separation of amoxicillin from pharmaceutical wastewater." *Iran Journal of Environmental Health Science & Engineering* 8 (2):109–116.

Shannon, M. A., P. W. Bohn, M. Elimelech, J. G. Georgiadis, B. J. Mariñas, and A. M. Mayes. 2008. "Science and technology for water purification in the coming decades." *Nature* 452:301. doi:10.1038/nature06599.

Shantanu, S. https://www.steviashantanu.com/stevia-membrane-based-extraction.

Shuler, M.L., F. Kargi, and M. DeLisa. 2017. *Bioprocess Engineering: Basic Concepts*. New Delhi, India: Pearson Education.

Sparks, T., and G. Chase. 2013. *Filters and Filtration Handbook*. Oxford, UK: Elsevier Science & Technology.

Stanbury, P. F., A. Whitaker, and S. J. Hall. 2017. "Chapter 5—Sterilization." In *Principles of Fermentation Technology* (*Third Edition*), 273–333. Oxford, UK: Butterworth-Heinemann.

Starbard, N. 2008. *Beverage Industry Microfiltration*. New York: John Wiley & Sons.

Suárez, A., T. Fidalgo, and F. A. Riera. 2014. "Recovery of dairy industry wastewaters by reverse osmosis. Production of boiler water." *Separation and Purification Technology* 133:204–211. doi:10.1016/j.seppur.2014.06.041.

Sutherland, K. 2011. "Pharmaceuticals: Filtration plays key role in pharmaceuticals and biotechnology." *Filtration + Separation* 48 (2):16–19. doi:10.1016/S0015-1882(11)70080-5.

Tamime, A. Y. 2012. *Membrane Processing: Dairy and Beverage Applications*. New York: John Wiley & Sons.

Tang, M., S. C. Chen, D. Q. Chang, X. Xie, J. Sun, and D. Y. H. Pui. 2018. "Filtration efficiency and loading characteristics of PM2.5 through composite filter media consisting of commercial HVAC electret media and nanofiber layer." *Separation and Purification Technology* 198:137–145. doi:10.1016/j.seppur.2017.03.040.

Tien, C., and B. V. Ramarao. 2017. "Modeling the performance of cross-flow filtration based on particle adhesion." *Chemical Engineering Research and Design* 117:336–345. doi:10.1016/j.cherd.2016.09.011.

Tipnis, N. P., and D. J. Burgess. 2017. "Sterilization of implantable polymer-based medical devices: A review." *International Journal of Pharmaceutics*. doi:10.1016/j.ijpharm.2017.12.003.

Tiwari, S. S. K., W. P. Schmidt, J. Darby, Z. G. Kariuki, and W. J. Marion. 2009. "Intermittent slow sand filtration for preventing diarrhoea among children in Kenyan households using unimproved water sources: randomized controlled trial." *Tropical Medicine & International Health* 14 (11):1374–1382. doi:10.1111/j.1365-3156.2009.02381.x.

Toledano, A., A. García, I. Mondragon, and J. Labidi. 2010. "Lignin separation and fractionation by ultrafiltration." *Separation and Purification Technology* 71 (1):38–43. doi:10.1016/j.seppur.2009.10.024.

Udén, P. 2006. "Recovery of insoluble fibre fractions by filtration and centrifugation." *Animal Feed Science and Technology* 129 (3):316–328. doi:10.1016/j.anifeedsci.2006.01.011.

United States Department of Energy. 2015. "Specification for HEPA filters used by DOE contractors." In *Department of Energy Technical Standards*. United States Department of Energy.

Vourch, M., B. Balannec, B. Chaufer, and G. Dorange. 2008. "Treatment of dairy industry wastewater by reverse osmosis for water reuse." *Desalination* 219 (1):190–202. doi:10.1016/j.desal.2007.05.013.

Wang, Y. J., Y. F. Duan, L. G. Yang, Y. M. Jiang, C. J. Wu, Q. Wang, and X. H. Yang. 2008. "Comparison of mercury removal characteristic between fabric filter and electrostatic precipitators of coal-fired power plants." *Journal of Fuel Chemistry and Technology* 36 (1):23–29. doi:10.1016/S1872-5813(08)60009-2.

Wei, H., X. Zhang, X. Tian, and G. Wu. 2016. "Pharmaceutical applications of affinity-ultrafiltration mass spectrometry: Recent advances and future prospects." *Journal of Pharmaceutical and Biomedical Analysis* 131:444–453. doi:10.1016/j.jpba.2016.09.021.

Winans, J. D., K. J. P. Smith, T. R. Gaborski, J. A. Roussie, and J. L. McGrath. 2016. "Membrane capacity and fouling mechanisms for ultrathin nanomembranes in dead-end filtration." *Journal of Membrane Science* 499:282–289. doi:10.1016/j.memsci.2015.10.053.

Wu, Z., M. Khan, S. Mao, L. Lin, and J. M. Lin. 2018. "Combination of nano-material enrichment and dead-end filtration for uniform and rapid sample preparation in matrix-assisted laser desorption/ionization mass spectrometry." *Talanta* 181:217–223. doi:10.1016/j.talanta.2018.01.016.

Xie, X., C. Le Men, N. Dietrich, P. Schmitz, and L. Fillaudeau. 2018. "Local hydrodynamic investigation by PIV and CFD within a Dynamic filtration unit under laminar flow." *Separation and Purification Technology* 198:38–51. doi:10.1016/j.seppur.2017.04.009.

Yazgan-Birgi, P., M. I. H. Ali, and H. A. Arafat. 2018. "Estimation of liquid entry pressure in hydrophobic membranes using CFD tools." *Journal of Membrane Science* 552:68–76. doi:10.1016/j.memsci.2018.01.061.

Yu, B. F., Z. B. Hu, M. Liu, H. L. Yang, Q. X. Kong, and Y. H. Liu. 2009. "Review of research on air-conditioning systems and indoor air quality control for human health." *International Journal of Refrigeration* 32 (1):3–20. doi:10.1016/j.ijrefrig.2008.05.004.

Yun, K. M., A. B. Suryamas, F. Iskandar, L. Bao, H. Niinuma, and K. Okuyama. 2010. "Morphology optimization of polymer nanofiber for applications in aerosol particle filtration." *Separation and Purification Technology* 75 (3):340–345. doi:10.1016/j.seppur.2010.09.002.

Zhao, P., J. A. Siegel, and R. L. Corsi. 2007. "Ozone removal by HVAC filters." *Atmospheric Environment* 41 (15):3151–3160. doi:10.1016/j.atmosenv.2006.06.059.

4 Membrane-Based Separation Processes

Kit Wayne Chew, Bervyn Qin Chyuan Tan, Jiang Chier Bong, Kevin Qi Chong Hwang, and Pau Loke Show

4.1 INTRODUCTION

Membrane-based separation is one of the most widely used bioseparation methods in the industry now. It is used in food processing, water treatment, and blood fractionation (McGregor, 1986). Throughout the past two to three decades, new membranes, module designs, and systems have been introduced to meet the demands of the biotechnological industry. The modern membranes are usually specially designed to purify, formulate, produce, or remove a range of biomolecules such as proteins, viruses, DNA, or cell debris. Membrane filtration encompasses a large section of membrane-based separation, including microfiltration, ultrafiltration, and nanofiltration, which are similar in working principles but differ in both application and pore sizes. These filtration methods have different operation modes: normal flow and cross flow, also known as tangential flow. These techniques remove and separate biomolecules according to their sizes. These filtrations can be done using various types of modules such as hollow fiber, tubular, and spiral wound modules. Each module has its pros and cons that fit into different processes based on the demands. A relatively new technique that has emerged over the past decade or so is membrane chromatography. This technique segregates a mixture of biomolecules according to their affinity to the stationary phase, which is the membrane itself. Usually, to enhance the specificity of this method, functionalized ligands are attached covalently to the membrane's surface. Membrane chromatography's main objective is to reduce the mass transfer limitations present in packed bed chromatography. Apart from that, compared to packed bed, membranes may have better binding capacities for large biomolecules (Charcosset, 2012c).

This chapter describes the fundamentals of membrane-based separation processes, which include microfiltration, ultrafiltration, nanofiltration, high performance tangential flow filtration, and membrane chromatography. The types of membrane and modules used are also explained, along with examples of membrane support materials. The types of membrane modules reviewed are plate and frame module, spiral wound module, and hollow fiber module. Furthermore, the application of membrane filtration in biotechnology and fields outside biotechnology are illustrated.

4.2 MICROFILTRATION

Microfiltration is a pressure-driven separation process where the biological constituents with diameters between 0.1 and 10 μm are removed from the process fluid as they pass through the membrane (Baker, 2012). Figure 4.1 shows the various membrane processes and their specifications. The operating pressure for the microfiltration process is relatively low as it requires only pressure that is lower than 0.5 MPa. In the biological and pharmaceutical manufacturing industries, microfiltration usually works with other separation process such as ultrafiltration to ensure that the product stream is free from large contaminants. Generally, the microfiltration membrane can be divided into two principal types: screen filter and depth filter. Both filters can be identified based on their pore sizes on the membrane surface; the size of the pores in a screen filter is relatively smaller compared to those in a depth filter.

The small pore size of the screen filter blocks particles of larger sizes from passing through it. This causes the accumulation of retained particle on the surface, resulting in low performance of the membrane. The screen membrane filter is usually used for cross-flow microfiltration. This enables retained particles trapped on the membrane surfaces to be cleaned by the recirculating fluid. Unlike the screen filter, the large pore size of the depth filter allows the particles to enter the interior of the membrane and get adsorbed onto the pore wall. Depth filters have a larger

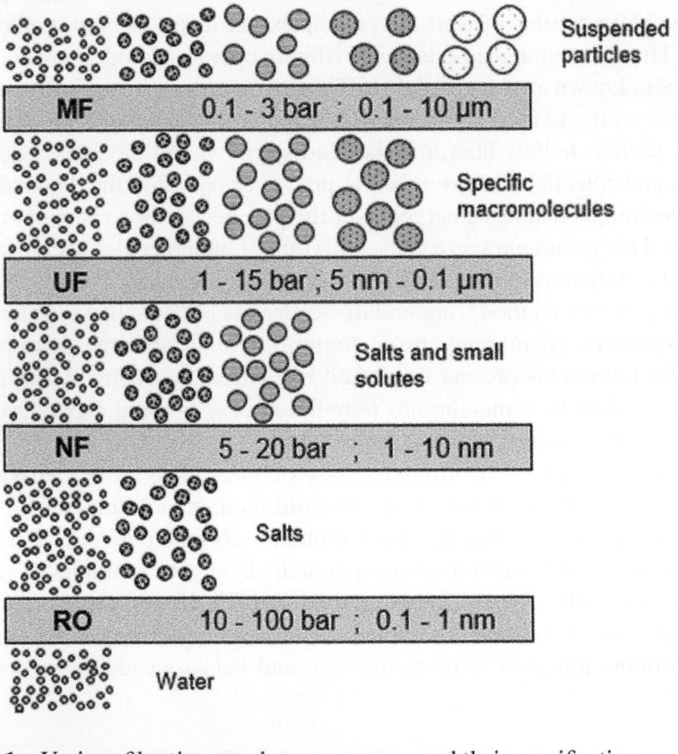

FIGURE 4.1 Various filtration membrane processes and their specifications.

surface area for the adsorption, giving them a larger holding capacity before fouling occurs. The adsorption mechanisms involved are Brownian diffusion, inertial capture, and electrostatic adsorption. Depth membrane filters are used for in-line filtration purposes due to their cost-effectiveness, especially in systems involving heavily fouling feed stocks.

During membrane manufacturing, completely filtering out bacteria is the criterion for selection of the membrane pore size. In a medium with low organism concentration, membranes with large pore size were found to be effective in bacteria removal. If there is an increase in concentration, the membrane may fail to quarantine all the bacteria. However, for membranes with small pore size, there will not be any breakthrough of bacteria, even in concentrated solutions. Therefore, after years of research on bacterial challenge tests, membranes with pore diameters of 0.22 μm and 0.45 μm are found to be more reliable for the filtering of bacteria (Baker, 2012). Microfiltration membranes have a lower energy requirement compared to ultrafiltration and nanofiltration membranes. They are also highly efficient in removing most bacteria and unwanted cell debris from the desired medium. Due to their outstanding performance, the lifespan of the subsequent separation equipment is prolonged, with lower maintenance needed.

Due to the different rates of permeation of constituents, however, there is a gradual build-up in concentration of nonpermeating components as more permeable components pass through the membrane, which results in layer formation on the membrane surface. This phenomenon is known as concentration polarization. Concentration polarization reduces the concentration difference of permeating components across the membrane, resulting in a subsequent decrease in the flux as well as in the selectivity of the membrane. Membrane fouling is also a major limitation that affects all the membrane technology performance, including microfiltration, as a result of the deposition of solutes or particles on the surface (external fouling) or in the membrane pores (internal fouling). Intense chemical cleaning or replacement of the membrane may be required if the fouling is severe. The pore size of a microfiltration membrane is usually larger than the molecule; thus, biomolecules like proteins are less likely to deposit at the membrane surface compared to other membrane application such as ultrafiltration membranes (Charcosset, 2012b).

Microfiltration membranes are commonly used for bacterial removal in various industries. The initial clarification of fermentation broths may be conducted by the microfiltration membrane to separate the suspended cell mass or debris such as proteins, antibiotics, lactic acid, and polysaccharides (Charcosset, 2012b). Microfiltration membranes are also used extensively for therapeutic and commercial purposes to separate plasma from blood (Basile and Charcosset, 2016). The technique of microfiltration and membrane adsorption is implemented under the same membrane filtration device to separate the human plasma proteins, albumin, and immunoglobulin. Apart from that, new techniques such as affinity microfiltration process is introduced to purify the immunoglobulin G and lactoferrin from cheese whey by combining affinity chromatography with microfiltration membrane (Basile and Charcosset, 2016). Adsorbents such as protein G-Sepharose, heparin-Sepharose, and protein-G-bearing streptococcal cells are used with microfiltration membranes of 0.2 μm in order to trap these macroligands.

4.3 ULTRAFILTRATION

Ultrafiltration is a type of membrane filtration using pressure forces or concentration gradients to separate colloids and macromolecules from water and microsolutes. As shown in Figure 4.1, the operating pressure for ultrafiltration process is relatively low because it works within the pressure range of 0.1–1.5 MPa (Grandison, 1996). The membrane pore size for ultrafiltration lies within the range of 5–100 nm, and this characteristic has made this technology widely used in recovering valuable constituents from fermentation broths and food-processing streams such as the purifying of protein solution. Ultrafiltration membranes have an anisotropic structure, which is a finely porous surface layer for separation and microporous substrate for mechanical support for the membrane (Ghosh et al., 2013). The capability of ultrafiltration membranes to retain macromolecules is usually referred to its molecular weight cut-off (MWCO). Although ultrafiltration membranes and microfiltration membranes have many similar characteristics in terms of their configuration, construction, and application, the types of membrane can still be selected based on their differences in pore size, which are designated to separate the materials with different sizes. Processing of dissolved macromolecules and separation of dispersed constituents such as the colloids, cells, or fat globules are conducted with ultrafiltration and microfiltration membrane, respectively (Grandison, 1996). The separation mechanism of an ultrafiltration membrane depends not only on the pore sizes but also considers the charge of the macromolecule as well as its affinity for the filtering membrane (Doelle et al., 2009).

Ultrafiltration has emerged as an alternative to concentrating protein and exchanging buffer, which is mainly done by size-exclusion chromatography for the purification of plasmid DNA or virus-like particles. The ultrafiltration membrane has drawn attention due to its thermal stability and inherent chemical stability. The ultrafiltration membrane is capable of separating macromolecules based on their charges and affinity (Doelle et al., 2009). Similar to the microfiltration membrane, the drawback of ultrafiltration membranes is the concentration polarization, which results in membrane fouling. Fouling tends to occur more often in ultrafiltration membranes compared to microfiltration membranes because the MWCO of ultrafiltration membranes that is smaller than the molecule weight of protein (Charcosset, 2012b).

In the biotechnology industry, ultrafiltration membranes are applied for protein separation. For examples, for both clarification of antibiotic broths (Boi, 2015) and purification of DNA (Li et al., 2016), ultrafiltration membranes are applied for the downstream process to recover the desired products. The extension of ultrafiltration, which is diafiltration, is used to concentrate the enzyme solution, namely, the lysozyme (Charcosset, 2012a). Ultrafiltration membranes are applied before the chromatographic steps as pretreatment of the feed stream. It can be implemented for the purpose of desalting, polishing, and buffer exchange. This combination of ultrafiltration membrane and chromatography is used to purify antibodies, DNA, and other biomolecular species such as the IgG and fibronectin from blood plasma. This adsorptive membrane chromatography is capable of maintaining its performances at high flow rates and preventing the denaturation and degradation of biomolecules (Basile and Charcosset, 2016).

4.4 NANOFILTRATION

Nanofiltration is a relatively new, pressure-driven membrane separation process generally used to partially demineralize and separate low-molecular-weight components such as small organics, monovalent ions, and salts from the larger organic compounds. The pressure involved in this method is within the range of 0.5–2 MPa (Basile and Charcosset, 2016). Nanofiltration membranes have pore sizes of 1–10 nm, and the MWCO of the membranes lies within the range of 100–1000 (Figure 4.1). The prospects of high divalent ion rejection have allowed nanofiltration membranes to be used extensively in many areas, especially in the pharmaceutical and biotechnology sectors. An asymmetric structure is usually found in the nanofiltration membrane, where there is a thin layer superimposed on a thicker sublayer that provides a porous support. The thin membrane takes charge of the separation process and resists the water flow that contains negatively charged particles (Bruggen and Geens, 2009). The separation mechanism of nanofiltration membranes is relatively complicated and is based on the microhydrodynamic and interfacial events occurring at the surface of the membrane and within the nanopores of the membrane.

A combination of steric, Donnan, dielectric, and transport effects influences the separation process (Agboola et al., 2015). The steric mechanism, also known as the size-based exclusion, is utilized to remove neutral solutes. The equilibrium and membrane potential interactions between charged particles and the interface of the charged membrane are usually explained with the Donnan effect. The differences in polarization charges at the dielectric boundary between pore walls and pore-filling solution can induced dielectric exclusion. The nanofiltration membrane can also remove ions through electrostatic interaction. To be more specific, the nanofiltration membrane removes dissolved solids that carry bivalent ions from the medium, which cannot be done by either the microfiltration membrane or the ultrafiltration membrane (Hung et al., 2012). By implementing the nanofiltration membrane, the volume of wastewater can be reduced as the filtrate can be directly discharged into sewers (Le et al., 2014).

Membrane fouling and concentration polarization are still the greatest challenge for all membrane-based separation techniques, and this includes nanofiltration membranes. Due to the membrane characteristics of surface hydrophobicity and permeability, the nanofiltration membrane tends to have fouling issues compared to other membrane separation techniques (Van der Bruggen et al., 2008). In addition, greater amounts of energy are required by the nanofiltration membrane to drive the purification process compared to the microfiltration or ultrafiltration membrane. The nanofiltration membrane is also incapable of removing monovalent ions, such as chloride and sodium ions, efficiently (Hung et al., 2012).

Nanofiltration membranes have been discovered to be suitable for enhancing viral safety in biopharmaceutical product manufacturing, such as inactivation of the hepatitis virus (Dichtelmüller et al., 2012). The membrane prevents the infectious agents from invading human plasma pools. It is used to isolate the antiviral drug precursor such as N-acetyl-d-neuraminic acid (Bowen et al., 2004). The membrane can be integrated with a membrane crystallizer to achieve zero liquid discharge for physicochemical wastewater treatment (Tun and Groth, 2011). It is proven that this technique can successfully remove more than 99% of sodium sulphate from aqueous wastes (Curcio et al., 2010). A nanofiltration-electrodialysis technique has also been introduced as a pretreatment process for various types of wastewater treatment. Nanofiltration membranes have been targeted for the removal of organic compounds and also partial desalination, while electrodialysis is used to achieve a higher degree of water purity. This technique is extensively used for the conversion of organic acids, such as lactic acid (Basile and Charcosset, 2016). The development of nanofiltration membranes incorporated with photocatalysis also increased the efficiency of the dye degradation process in wastewater treatment (Samhaber and Nguyen, 2014).

4.5 HIGH-PERFORMANCE TANGENTIAL FLOW FILTRATION

High-performance tangential flow filtration (HPTFF) is an efficient and emerging technology for the separation of biomolecules by using semipermeable membranes. HPTFF exploits the differences of sizes and charges that are characteristic of biomolecules for the separation process without limit to their protein relative size (Van Reis et al., 1997). The biomolecules are separated with proper optimization of buffer and fluid dynamics. Depending on the membrane porosity, the separation process can be classified as microfiltration or ultrafiltration. HPTFF has been used to separate monomers from oligomers based on their differences in size, protein variants differing at only a single amino acid residue, and an antigen binding fragment from a similar size impurity. This technology can affect simultaneous purification, concentration, and buffer exchange, providing an opportunity to combine several different separation steps into a single scalable unit operation. The high selectivity in HPTFF processes is obtained by exploiting a number of different phenomena that are the control of device fluid mechanism and filtrate flux, which are able to minimize the fouling effect in the system and exploit the concentration polarization effect. Figure 4.2 illustrates a simple HPTFF system. The working principle is that the fluid is pumped from the reservoir across the membrane surface. Next, the

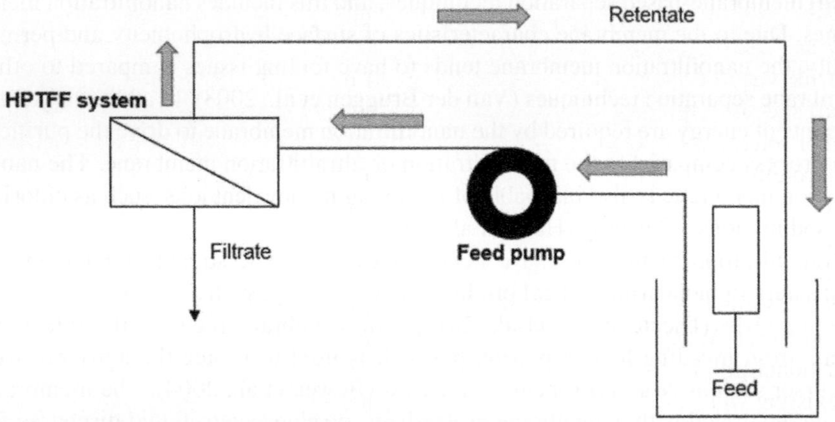

FIGURE 4.2 The working principle of high-performance tangential flow filtration system.

retentate exits the HPTFF system and returns to the sample reservoir. The filtrate exiting the HPTFF is collected at the filtrate collection chamber.

Common technologies such as ion-exchange chromatography, ultrafiltration, and size exclusion chromatography are often used for buffer exchange, concentration, and purification processes. Compared to all these techniques, HPTFF technology can provide high-resolution purification while maintaining the inherent high throughput and high yield characteristics of ultrafiltration technology. Therefore, HPTFF can perform in initial, intermediate, and final purification stages. In tangential flow filtration, the solution with solutes is pumped tangentially along the surface of the semi-permeable membrane. The high selectivity of this technology is given by the pressure-dependent regime where pressure is applied, which forces the fluid through the membrane to the filtrate side. Pressure drop is unfavorable in the separation process, so the high selectivity is done by maintaining the transmembrane pressure throughout the module. One of the simplest approaches for minimizing the transmembrane pressure variation in the system is to establish a co-current flow on the filtrate side of the semi-permeable membrane by using a recirculation pump to generate a desire pressure gradient in the filtrate channel. This ensures that the gradient is balanced in the retentate throughout the separation process (Van Reis et al., 1997). The flux is influenced by the tangential flow velocity, which is a measure of shear turbulence in the flow channels. However, the shear force is highly dependent on several factors such as viscosity and density of the retentate, as well as the flow channel diameter. These factors vary over the duration of filtration process. For most of the application, the flow increases with tangential flow velocity to ensure that the gel polarization layer is minimized within the constraints imposed by the minimum allowable pressure drop throughout the filtration process.

The performance of HPTFF can be improved by controlling the buffer pH and the ionic strength. This greatly maximizes differences in the effective volume of the desired product and impurities. This can be done by increasing the protein charge or reducing the ionic strength of the solution, which in turn increases the effective volume and reduces the protein transmission through the membrane. In addition, the electrostatic exclusion of the additional retained species is maximized by operating the system close to the isoelectric point of the lower weight protein and at a relatively low salt concentration (Christy et al., 2002). The overall performance of HPTFF can likewise be improved by using a charged membrane to increase the retention of all species with like-polarity. A sieved positively charged membrane can provide higher retention of positively charged protein. These repulsive interactions prevent the proteins from entering the membrane, resulting in lower protein transmission and reduced membrane fouling (Christy et al., 2002).

Certain applications for HPTFF include concentration, desalting, and buffer exchange, and fractionation of protein. HPTFF can be combined with various kind of processes, for example, purification, concentration, and formulation, into a single unit for operation. Whey is a relatively inexpensive source of valuable proteins that contain nutritional properties. However, the unique properties in whey is always unrealized in existing whey products due to the adverse interactions between components when conventional separation methods are applied. HPTFF can purify the complex protein mixtures in whey inexpensively. This is achieved by the control of membrane range, transmembrane pressure, desalting and buffer exchange, solution pH and ionic strength. A hollow fiber module containing a microfiltration membrane was used with HPTFF to process the recombinant *E. coli* lysates, which contain protein inclusion (Bailey and Meagher, 1997). Significant differences in the permeate flux and protein transmission were observed for various membranes compared to the best performing membrane. Since the operating condition of the homogenization process greatly affects the filtration process of *E. coli*, the suitable condition of homogenization was carefully studied. There have also been studies on the relationship between increased homogenization and the centrifugal recovery of soluble protein from yeast (Clarkson et al., 1993) and inclusion bodies from *E. coli* lysate (Titchener–Hooker et al., 1991).

HPTFF exhibits a number of advantages over many applications in term of separation process. It can recover most of the biomolecules such as antibodies and recombinant protein (from *E. coli*). HPTFF is mostly applied in biotechnology fields for the concentration and desalting of proteins and peptides. It is also applicable on sample preparations prior to column chromatography for protein purification processes. Nevertheless, HPTFF also has certain drawbacks. It cannot be operated at high transmembrane pressure because it affects product purity. The recovery of proteins can also be limited due to the concentration polarization effect. This can be improved by enhancing the resolution of the protein, where a charged membrane can be used for the positively charged membrane to repel similar charged proteins. This results in a like-charge repulsion, preventing the protein from entering the membranes and thus lowering the protein transmissions (Christy et al., 2002).

4.6 MEMBRANE CHROMATOGRAPHY

Chromatography has been widely used for purification or analytical purposes (Hostettmann et al., 1986). The mechanism of chromatography is separation of a mixture via partitioning. The mixture dissolved in a fluid called the mobile phase is passed through a material called the stationary phase whereby each component of the mixture travels at varying speeds, resulting in their separation. This separation depends on the partition coefficient of each component (McMurry, 2014). Membrane chromatography usually utilizes microfiltration techniques with added functional ligands affixed to the internal pores' surfaces. This allows for highly selective separations via adsorption/binding interactions. However, affinity-based, ion exchange, hydrophobic, and reversed phase membranes have also been developed. There are a number of advantages of membrane chromatography in comparison with typical chromatography. The main setbacks that membrane chromatography can overcome are high mass transfer resistance, high pressure drops, low flow rates, and long processing times. Membrane chromatography has a wide range of applications, including protein purification, protein analysis, polishing, plasmid DNA purification, virus purification, and monoclonal antibody purification (Orr et al., 2013).

A typical membrane chromatographic system is comprised of one or more microporous membranes as the stationary phase. These membranes carry adsorptive functional groups with them that are based on ion exchange, affinity, immunoaffinity, or hydrophobic interactions that are covalently bonded on the membranes. Each of these individual membranes in the system can have different geometries, from disks or layered sheets to hollow or full fibers (Pabby et al., 2015). Affinity-based chromatography is considered to cover the largest portion of the membrane chromatography segment and is also one of the most selective methods for purification (Charcosset, 2012c). Ligands in affinity membranes are immobilized on the surface of membranes. As the feed mixture passes through, desired molecules bind selectively and reversibly to the ligands. Examples that use this force of attraction are antigens and antibodies, and enzymes and substrates. These ligands are connected to the membrane by a spacer molecule to allow a larger range of orientations of the ligand, which increases the formation of ligate-ligand complexes. Table 4.1 shows the examples of functionalized groups or ligands and their corresponding ligates along with the type of chromatography involved.

As reported by Zhu et al. (2011), poly(vinyl alcohol-co-ethylene) (PVA-co-PE) membranes with Cibacron Blue F3GA (CB) incorporated on the surface had a very specific adsorption of BSA. The membrane had a high Bovine Serum Albumin (BSA) capture of 105.5 mg/g of membrane and nonspecific BSA binding to the pristine nanofibrous membrane was below 4 mg/g membrane. Moreover, the attachment of ligands on the membrane reached 220 mg/g, which is very high. The elution efficiency achieved 92.5%. Even after ten adsorption-elution cycles, the capture capacity dropped only slightly. This shows that this general affinity ligand (CB) immobilized on PVA-co-PE is feasible in biomolecular purifications (Zhu et al., 2011). Jiang et al. (2018) also published findings on immobilized metal ion affinity chromatography immobilized metal ion affinity chromatography (IMAC) with various metal

TABLE 4.1
Examples of Successful Ligand-Ligate Pairs in Protein Separation

Ligate Type	Type of Chromatography	Ligand Type	References
Bovine serum albumin (BSA)	Affinity	Cibacron Blue F3GA	Zhu et al. (2011)
Human immunoglobulin G (IgG)	Affinity	Protein A	Boi et al. (2008)
L-Zagreb mumps virus	Immunoaffinity	Polyclonal antibodies	Brgles et al. (2016)
Arginine-rich proteins	Affinity	Diethyl-4-aminobenzyl phosphonate (D4ABP)	Chenette et al. (2017)
Bovine serum albumin (BSA) and human immunoglobulin G (IgG)	Affinity	Poly N-vinylcaprolactam (PVCL)	Himstedt et al. (2013)
Phosphopeptides	Ion-exchange	Fe_3O_4@PDA-Nb^{5+} and Fe_3O_4@PDA-Ti^{4+}	Jiang et al. (2018)
Factor VII	Affinity	CNBr-activated Sepharose 4B	Hosseini and Nasiri (2015)
Histidine tagged proteins	Affinity	Nickel(II)-immobilized sulfhydryl cotton fiber (SCF-Ni^{2+})	He et al. (2015)

ions immobilized on polydopamine (PDA)-coated Fe_3O_4 (Fe_3O_4@PDA-M^{n+}). The research showed that ions of niobium (Nb^{5+}), titanium (Ti^{4+}), gallium (Ga^{3+}), and zirconium (Zr^{4+}) displayed superior enrichment efficiency. However, considering the sensitivity, Nb^{5+}, Ti^{4+} and Ce^{4+} immobilized Fe_3O_4@PDA might have an advantage over other metal ions. Besides, the results indicated that varying ions differ widely in enriching phosphopeptides, and Nb^{5+} and Ti^{4+} displayed greater performance in capturing phosphopeptides with both elevated selectivity along with better sensitivity. Furthermore, Nb^{5+} and Ti^{4+} showed significant advantages when used in real protein complex examples (nonfat milk) (Jiang et al., 2018).

Ion-exchange chromatography is another widely used method in separation and purification of biomolecules (Roberts et al., 2009). Biomolecules are separated based on their ionic charge. After the desired charged molecules are adsorbed on the membrane supports, they are eluted or, in simpler terms, removed from the stationary phase using a buffer with stronger ionic strength. The migration rate of biomolecules along the stationary phase is highly dependent on their charge density. As a high-concentration salt solution is used in elution containing the desired product, a diafiltration process is required to remove the salt and recover the product (Charcosset, 2012c). Table 4.2 shows the details of common ion exchange membranes (Han et al., 2005).

Teepakorn et al. (2015) studied the separation of two similar sized proteins called lactoferrin (LF) and bovine serum albumin (BSA) with both anionic and cationic exchange membrane chromatography. LF was completely retained when cationic membrane was used, while BSA was completely retained when an anionic membrane was used. The optimized pH is 6.0 for retention of both proteins. The selectivity of LF and BSA with both anionic and cationic exchange membranes were

TABLE 4.2
Types of Ion-Exchange Membrane with Details and Applications

Type of Ion Exchange Membrane*	pK_a	Functional Group (Ionic Ligands)	Membrane Charge	Example of Successful Applications	References
Q (strong anion)	11.0	$-CH_2-N^+-(CH_3)_3$	Positive	Recombinant NbZFP1 protein and antigens from mites extract recovery	Bartley et al. (2015), Wu et al. (2014)
S (strong cation)	1	$-CH_2-SO_3^-$	Negative	Purification of RNAi effector nuclease	Hammond et al. (2001)
D (weak anion)	9.5	$-CH_2-N^+-(C_2H_5)_2$	Positive	Removal of Cr^{6+} ions from water	Gode and Pehlivan (2005)
C (weak cation)	4.5	$-COO^-$	Negative	Purification of recombinant human interferon gamma	Wei et al. (2006)

* Q: quarternary ammonium groups, S: sulfonic acid groups, D: diethylamine groups, C: carboxylic acid groups, RNAi: ribonucleic acid interference.

not dependent on flow rate (>12.0 and <24.0 BV/min). Additionally, there was no significant drop in selectivity when LF concentration loads were increased (Teepakorn et al., 2015). Hydrophobic interaction membrane chromatography is among the least widely used mode in the purification of biomolecules (Roberts et al., 2009). The working principles of this technique are passing the mobile phase over the stationary phase immobilized with hydrophobic interaction ligands; this causes hydrophobic species in the mobile phase to interact with the said ligands. This method is mainly utilized in the separation of viruses, cells, nucleic acids, and carbohydrates. Elution is then conducted in rising levels of hydrophobicity. Ligands used in this method are at linear. aliphatic chains ranging from four to ten carbon atoms with a terminal amino group (Pabby et al., 2015). The membrane chromatography is not limited to a single type. Combinations of multiple membrane chromatographic supports are usually required to attain acceptable, higher levels of selectivity. Thus, a mix of hydrophobic interaction, ion exchange, reversed-phase, and affinity chromatography has been put forward for a myriad of different purification schemes.

4.7 MEMBRANES AND MODULES

4.7.1 Plate-Frame Module

The plate-frame module is fabricated with a flat-sheet membrane where there is flow channel exhibited as the space between the flat membrane plates. The permeates are obtained from the inside of the flat membrane plate. Figure 4.3 is a schematic illustration of a plate and frame module. There are two types of flat membrane modules: stack-type and plate-frame type. The stack-type module exhibits high membrane layer density by horizontally piling up membranes. This type of module is usually operated by tangential flow filtration technique (Camacho et al., 2013; Uragami, 2017). The plate-frame type module is applicable for feed with high concentrations of suspended material. Currently, a new design of plate-frame module has been

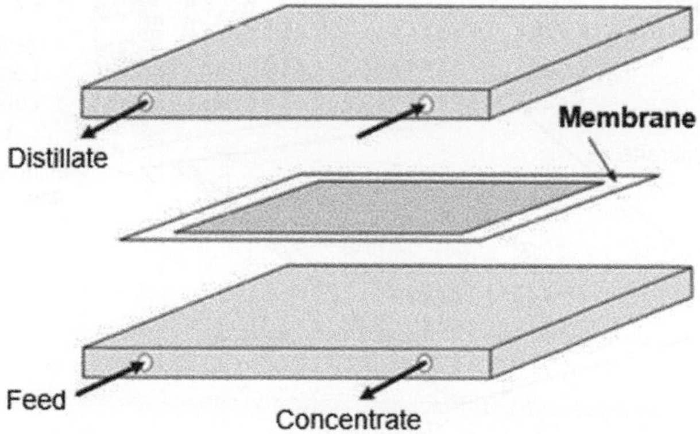

FIGURE 4.3 Schematic illustration of plate and frame module. (Camacho, L.M. et al., *Water*, 5, 94–196, 2013.)

introduced with multiple axial flow channels in place of sheet flows, leading to a better wall shear rate and average axial velocity. The major advantage of this invention is the low volume holdup within the channel height to process high viscosity of material.

4.7.2 Spiral-Wound Module

A spiral-wound module is composed of tightly packed filter media where porous plastic screen supports are wrapped around a flat-sheet permeable membrane. The edge of the permeable membrane is sealed and gapped with a spacer material, while the center consists of perforated tubes to allow the flow of liquid. Figure 4.4 is a schematic illustration of a spiral-wound module. Backpressure is the driving force, and filtration occurs inside the module. The impurities to be filtered flow through the membrane surface and spiral radially toward the weep holes in the central tubular tube, and the permeate is collected at the other end of the module. The major drawback of this design is the small dimension of the central flow channel, which could possibly increase the chance of plugging (Uragami, 2017).

4.7.3 Hollow-Fiber Modules

A hollow-fiber module is composed of a bundle of hollow fibers in either a straight configuration, which allows feed flow on the inside, or in a U-shape

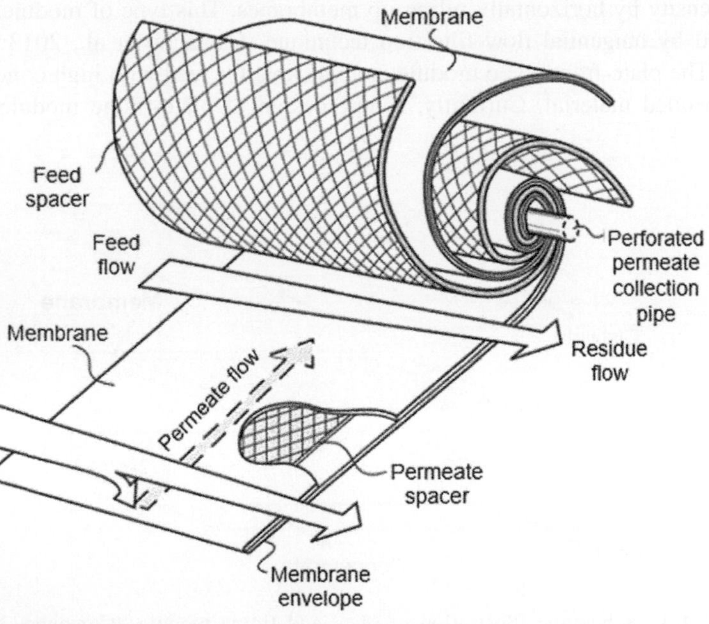

FIGURE 4.4 Schematic illustration of spiral-wound module.

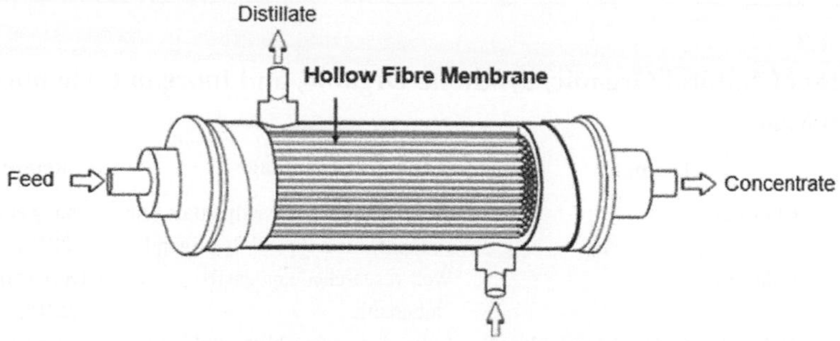

FIGURE 4.5 Schematic illustration of a shell and tube hollow-fiber module. (Camacho, L.M. et al., *Water*, 5, 94–196, 2013.)

configuration, allowing feed flow on the outside (Uragami, 2017). Figure 4.5 is a schematic illustration of a shell and tube hollow-fiber module. The edges of the membranes are stuck into the tube sheet to ensure that the membranes are not blocked during separation processes. A material of flow is not required in this module as the hollow fibers work for both separation and support (Belfort, 1988; Camacho et al., 2013).

Pressure is the driving force of the feed flow on the outside of the hollow fibers, and the permeates flow radially within through the unsupported membranes. Thus, the desired product flows inside the membranes before reaching the collection chamber.

4.8 MEMBRANE SUPPORT MATERIALS

Chromatographic membranes support materials, which are the stationary phases, are made from a wide variety of materials, ranging from natural organic, synthetic organic, and inorganic. Some examples of materials in each category are listed in Table 4.3.

Certain criteria must be met for a membrane chromatographic support to be successfully applied: It must (1) be microporous for large surface area of contact and free interaction of biomolecules with ligands, (2) be hydrophilic and neutrally charged to prevent nonspecific binding on the membrane, (3) consist of functional groups to allow activation via numerous reactions, (4) have high chemical stability to tolerate harsh environments during the entire chromatographic process, (5) be physically durable to withstand high trans-membrane pressure and sterilization by autoclaving, and (6) be low cost for industrial applications (Charcosset, 2012c).

Membrane chromatography that are equipped with functionalized ligands ceases with the need of diffusivity from bulk mixture to the adsorption sites. The diffusion process is very slow as compared to the convective flow involved in membrane chromatography. Theoretically, membrane chromatography should exceed efficiencies of conventional chromatography by factors of ten or higher. However, certain

TABLE 4.3
Examples of Natural Organic, Synthetic Organic, and Inorganic Membrane Support Materials

Category	Examples	Advantage	References
Natural organic	Chitosan	Novel, cheap, and easily attainable support for a typical dye ligand	Zhang et al. (2010)
	Cellulose	Well-researched and easily attainable	Teepakorn et al. (2015)
Synthetic organic	Nylon/poly(2-(methacryloyloxy) ethyl succinate)	Enhanced permeability and better protein capture rate	Anuraj et al. (2012)
	Poly(vinyl alcohol-co-ethylene)	Good protein capture and elution efficiency	Zhu et al. (2011)
	Polyvinylidene fluoride	Physically and chemically stable, biodegradable and cheap	Fan et al. (2017a, 2017b)
Inorganic	Alumina composite	Low nonspecific adsorption, excellent biocompatibility, easily functionalized	Shi et al. (2010)
	Aluminum oxide/chitosan	Easily regenerated, high elution efficiency, and high adsorption capacity	Shi et al. (2010)

anomalies in mass transfer effects slowing down the process, such as dead spaces and nonuniform flow, have been observed (Pabby et al., 2015). With regard to technical issues, ligands available currently are largely from existing molecules rather than being specifically designed, or they are ligates that are marked to have interactions with current ligands such as histidine tagging. Besides, specificity of most current ligands cannot be tuned sufficiently to meet requirements, while naturally specific ligands such as antibodies are unstable and expensive. The problems associated with stability of immunoaffinity ligands are the harsh environment during the elution process. Brgles et al. (2016) managed to utilize amino acid buffers to elute mumps virus at neutral pH, which is favorable for sensitive ligands (Brgles et al., 2016). Immunoaffinity ligands can cost upward of a few hundred dollars per milligram (Kress-Rogers, 1996), and their sensitivity and fragility prevents multiple uses. This causes capital cost to increase due to the mandatory oversizing of chromatography columns (Roy and Gupta, 2006). Membrane chromatography in general has lower binding capacity compared to their packed bed counterparts. To solve this issue, the membrane needs to have higher specific area, but this affects other properties such as permeability, pore size distribution, and mechanical strength. The most widely studied solutions to this issue are spacer arms, membrane supports, and pore coatings. The pore coatings create a three-dimensional area to increase binding surface (Ghosh, 2002).

4.9 APPLICATION OF MEMBRANE FILTRATION

4.9.1 Membrane Filtration in Biotechnology

Microfiltration membrane is widely used in the pharmaceutical industry for cold sterilization purposes to overcome the limitations using conventional method of sterilization with heat, for example, the loss in the effectiveness of a drug. Generally, a microfiltration membrane is employed to remove undesired suspensions and bacteria. A 0.2-μm-rated filter is commonly used in the pharmaceutical industry due to its effectiveness by providing a 10^7 reduction in bacteria (Baker, 2012). In biotechnology industries, during the recovery of intracellular products, microfiltration is used for cell debris removal and cell harvesting. The purification of bovine growth hormone from *Escherichia coli* lysate utilized a 0.1 μm microporous membrane for effective purification (Tutunjian, 1985). Microfiltration membranes with pore sizes of 0.3–0.5 μm also work well in separating the protease from fermentation broth.

Ultrafiltration is used extensively in the separation of protein through concentrating and desalting the protein by removal of water, salts, and other low-molecular-weight compounds. Chemicals such as ammonium sulphate, sodium sulphate, and caprylic acid can be removed from the protein with this technique (Ghosh et al., 2013). Ultrafiltration membranes are not usually used for clarification compared to microfiltration membranes because microfiltration membranes are sufficiently effective for trapping most of the bacteria. However, it is less possible for microfiltration membranes to trap viruses during pharmaceutical products manufacturing. For these cases, ultrafiltration membranes are the preferred application (DiLeo et al., 1992). The potential of using ultrafiltration membranes for separating DNA plasmid has also been researched in other studies (Borujeni and Zydney, 2014).

Diafiltration is an extension of ultrafiltration and utilizes water during the concentration process. Diafiltration is used extensively in the pharmaceutical biotechnology fields, especially in the removal of low-molecular-weight components such as cations, anions, small proteins, or other anti-nutritional compounds. Diafiltration is also employed for concentrating the enzyme solution. The exceptional properties of diafiltration allow it to be utilized as an alternative process for dialysis and ion exchange. The membrane is also employed in the medical field for hemodialysis. The purification of nucleic acids can also be conducted by ultrafiltration membranes (Li et al., 2016).

For the enhancement of viral safety in biopharmaceutical product manufacturing, nanofiltration membranes are a more convincing technique compared to the conventional viral reduction treatment using heat or solvent-detergent. Nanofiltration membranes are implemented to inactivate hepatitis virus and even the human immunodeficiency virus (Dichtelmüller et al., 2012). They can provide defense against infectious agents from potentially invading the human plasma pool. The capability of removing prions has made the membrane popular in biopharmaceutical industries (Burnouf and Radosevich, 2003). Other pharmaceutical applications employ nanofiltration membranes, such as the isolation of N-acetyl-d-neuraminic acid, which is an antiviral drug precursor (Bowen et al., 2004). Clindamycin can also be recovered from fermentation wastewater with the implementation of nanofiltration membranes (Zhu et al., 2003).

4.9.2 Virus Filtration

Viruses are essentially made of DNA or RNA and are either encapsulated by proteins, carbohydrates, or lipids or are not enclosed at all. Viruses are usually between 20 and 150 nm in size. Viral contamination can affect a variety of material, such as raw materials, cell cultures, and bioreactors. For these reasons, pharmaceutical companies especially have to practice viral safety and include virus clearance in their manufacturing processes. However, virus clearance is tedious and almost impossible due to complexities in testing for all possible viruses. The idea is to remove viruses to the point where the probability of viral contamination is extremely low. Both regulatory agencies from the United States and the European Union have set a risk minimization level of viral contamination in bioproducts derived from cell lines over the past 20 years (Marques et al., 2009). Virus clearance can be done in two ways: viral inactivation, which renders the virus useless but still present, or virus removal.

Virus filtration is one of the robust, size exclusion, virus removal techniques that can work in sequence with other virus clearance steps (Brorson et al., 2004). Because viruses are so small that minor defects in filters can lead to excessive virus leakage, manufacturing must be done to eliminate all possible defects (Van Reis and Zydney, 2007). Additionally, multiple layers of filter membranes are utilized as a contingency if any prior membranes fail. Virus filters were previously designed with the intention for use only in tangential flow filtration (TFF) to reduce the propensity of fouling. However, the simple design and low cost of normal flow filtration (NFF) has led to the extensive use of virus filters specifically intended for NFF. Compared to TFF, these normal flow filters are typically utilized with the larger pore size of the membrane facing the feed stream. This allows for the larger foulants such as protein aggregates to be retained within the macropores first. This in turn protects the sensitive virus retention layer. Single-use normal flow filters have also been introduced to further simplify system design while reducing labor, operating, and capital costs.

Some of the common materials used to manufacture virus filtration membranes are hydrophilic polyether-sulfone, cellulose, and polyvinylidene fluoride (PVDF). Some of the commercially available virus filters are listed in Table 4.4.

As with all other filtration processes, fouling in virus filtration is fairly common and poses a significant problem. Among the sources of fouling are protein aggregates, denatured products, DNA, and random debris. The most widely researched solution to this problem is the use of prefilters. Prefilters are essentially guard filters that remove damaging foulants within the feed stream to enhance the robustness, performance, and capacity of the virus filter (Bolton et al., 2010). Alternatively, ion exchange membranes have been successfully utilized to improve the filtration of parvovirus and various antibodies (Brown et al., 2010). By using this technique, a relatively clean and chemically stable environment can be provided to enhance throughput.

On another note, the freezing of a material to be filtered might have an adverse effect on the filtration performance. One study showed that filtering fresh feed used up to six times less filter area compared to thawed frozen feed. A prefilter can be used to reverse the detrimental effect of freezing the feed (Ireland et al., 2006).

TABLE 4.4
Examples of Commercially Available Virus Filters

Company	Filter Model	Type of Filter	Type of Virus	Virus Size (nm)
Asahi-Kasei	Planova 15N	Hollow regenerated cellulose fiber	Parvovirus and poliovirus	18–30
	Planova 20N		Parvovirus and encephalomyocarditis	18–30
	Planova 35N		Human immunodeficiency virus (HIV) and bovine viral diarrhea virus	40–130
Millipore	Viresolve NFP	Composite of ultrafiltration and microfiltration	ϕX-174 bacteriophage	>28
	Viresolve NFR		Retrovirus	80–130
Pall	ULtipor DV20	Hydrophilized PVDF	PP7 bacteriophage	>26
	ULtipor DV50		PR772 bacteriophage	76–88

Source: Kern, G. and Krishnan, M., *Biopharm. Int.*, 19, 2006.

4.9.3 Membrane Filtration outside Biotechnology

Microfiltration membranes are also used in the food industry with the same purpose of cold sterilization to overcome the limitations of sterilization with heat, which include losing flavor and the nutrient value of foods. However, unlike the pharmaceutical industry, in the food industry, including wine manufacturing, the membrane requirement is much less strict because the main target is just to remove the leftover yeast cells. Thus, a filter of 1 μm pore size is of utmost significance in removing all the yeast along with 10^6 reduction in bacteria within the drink (Baker, 2012). Other than that, microfiltration membranes are also used in the dairy industry, especially in the production of milk and whey. In order to prolong the shelf life of the product, pasteurization using microfiltration membranes is conducted by prohibiting bacteria and associated spores from passing through the membrane. The great efficiency of 1.8 μm microporous membranes on the separation of casein from whey protein has made microfiltration membranes a vital technology in this industry. This is because the stream that is rich in casein is sent for cheese manufacturing, while the other whey protein stream is further processed into whey protein concentrate (WPC) and whey protein isolate (WPI) powders (Hansen, 1988). Microfiltration membranes are also used in the application of water treatment, where they act as the first boundary of disinfection for the uptake water stream. Microfiltration membranes remove pathogens such as protozoa Cryptosporidium and Giardia lamblia (Starbard, 2009). Microfiltration membranes work well in treating the turbidity of secondary wastewater effluents (Bhattacharya et al., 2013).

Ultrafiltration membranes have several applications in the food industry because of their great ability to recover protein molecules. Ultrafiltration membranes are widely used in the dairy industry, especially in the manufacturing of cheese whey. The membrane can produce WPC that is 10 to 30 times more concentrated than the feed

(Cheryan, 1998). The reduction of lactose and calcium in milk is performed using ultrafiltration membranes (Kumar et al., 2013). Ultrafiltration is also considered as more cost effective and energy efficient as no heating process is required. The removal of particulates and macromolecules from raw water can be performed by ultrafiltration membranes. This filtration technique acts as an alternative for secondary and tertiary filtration systems (Clever et al., 2000). Apart from that, ultrafiltration membranes are employed as an advance tertiary treatment process for wastewater (Falsanisi et al., 2009).

Nanofiltration is considered an alternative to the conventional ways of food processing, especially in the dairy industry. Nanofiltration membranes are utilized to improve the recovery of lactose by removing minerals such as potassium, calcium, and sodium from the sweet whey and skim milk permeates. As a result, nanofiltration plays a great role in increasing the heat stability of milk (Guu and Zall, 1992). With the implementation of nanofiltration membranes, the volume of wastewater is reduced because the nanofiltrate can be discharged directly into sewers (Le et al., 2014). Nanofiltration filtration has proven to be a great technique on water softening, which requires the removal of hardness ions, such as magnesium and sodium, that are bivalent (Nanda et al., 2008).

4.10 CONCLUSION

This chapter explained the fundamentals of membrane-based separation and modules, which includes the principles and technologies used in membrane processes. Various types of membrane functionality and potential benefits and drawbacks were explained. These membranes are very effective in biomolecule separation; however, the membrane are prone to fouling, which is a challenge in industries for membrane maintenance. The modification of operating conditions may help to prevent membrane fouling. As an alternative to the physical techniques, chemical cleaning can also be used to restore filter performance. The membrane modules and supports used in membranes were described in this chapter. The application of these membranes in biotechnology and other fields was illustrated. The developments of better membrane material performance and lower driving force operating conditions will be required to improve the energy efficiency of the membrane processes.

REFERENCES

Agboola, O., J. Maree, A. Kolesnikov, R. Mbaya, and R. Sadiku. 2015. "Theoretical performance of nanofiltration membranes for wastewater treatment." *Environmental Chemistry Letters* 13 (1): 37–47. doi:10.1007/s10311-014-0486-y.

Anuraj, N., S. Bhattacharjee, J. H. Geiger, G. L. Baker, and M. L. Bruening. 2012. "An all-aqueous route to polymer brush-modified membranes with remarkable permeabilites and protein capture rates." *Journal of Membrane Science* 389: 117–125. doi:10.1016/j.memsci.2011.10.022.

Bailey, S. M. and M. M. Meagher. 1997. "Crossflow microfiltration of recombinant Escherichia coli lysates after high pressure homogenization." *Biotechnology and Bioengineering* 56 (3): 304–310. doi:10.1002/(SICI)1097-0290(19971105)56:3<304:AID-BIT8>3.0.CO;2-N.

Baker, R. W. 2012. "Microfiltration." In *Membrane Technology and Applications*, pp. 303–324. Chichester, UK: John Wiley & Sons.

Bartley, K., H. W. Wright, J. F. Huntley, E. D. T. Manson, N. F. Inglis, K. McLean, M. Nath, Y. Bartley, and A. J. Nisbet. 2015. "Identification and evaluation of vaccine candidate antigens from the poultry red mite (Dermanyssus gallinae)." *International Journal for Parasitology* 45 (13): 819–830. doi:10.1016/j.ijpara.2015.07.004.

Basile, A. and C. Charcosset. 2016. *Integrated Membrane Systems and Processes*, Chichester, UK: John Wiley & Sons.

Belfort, G.. 1988. "Membrane modules: Comparison of different configurations using fluid mechanics." *Journal of Membrane Science* 35 (3): 245–270. doi:10.1016/S0376-7388(00)80299-9.

Bhattacharya, P., A. Roy, S. Sarkar, S. Ghosh, S. Majumdar, S. Chakraborty, S. Mandal, A. Mukhopadhyay, and S. Bandyopadhyay. 2013. "Combination technology of ceramic microfiltration and reverse osmosis for tannery wastewater recovery." *Water Resources and Industry* 3: 48–62.

Boi, C., S. Dimartino, and G. C. Sarti. 2008. "Performance of a new protein a affinity membrane for the primary recovery of antibodies." *Biotechnology Progress* 24 (3): 640–647. doi:10.1021/bp0704743.

Boi, C.. 2015. "Recovery of antibiotics from fermentation broth membrane operations." In *Encyclopedia of Membranes*, edited by Enrico Drioli and Lidietta Giorno, pp. 1–2. Berlin, Germany: Springer.

Bolton, G. R., J. Basha, and D. P. LaCasse. 2010. "Achieving high mass-throughput of therapeutic proteins through parvovirus retentive filters." *Biotechnology Progress* 26 (6): 1671–1677.

Borujeni, E. E. and A. L. Zydney. 2014. "Membrane fouling during ultrafiltration of plasmid DNA through semipermeable membranes." *Journal of Membrane Science* 450: 189–196.

Bowen, W. R., B. Cassey, P. Jones, and D. L. Oatley. 2004. "Modelling the performance of membrane nanofiltration—Application to an industrially relevant separation." *Journal of Membrane Science* 242 (1): 211–220. doi:10.1016/j.memsci.2004.04.028.

Brgles, M., D. Sviben, D. Forčić, and B. Halassy. 2016. "Nonspecific native elution of proteins and mumps virus in immunoaffinity chromatography." *Journal of Chromatography A* 1447: 107–114.

Brorson, K., L. Norling, E. Hamilton, S. Lute, K. Lee, S. Curtis, and Y. Xu. 2004. "Current and future approaches to ensure the viral safety of biopharmaceuticals." *Developments in Biologicals* 118: 17.

Brown, A., C. Bechtel, J. Bill, H. Liu, J. Liu, D. McDonald, S. Pai, A. Radhamohan, R. Renslow, and B. Thayer. 2010. "Increasing parvovirus filter throughput of monoclonal antibodies using ion exchange membrane adsorptive pre-filtration." *Biotechnology and Bioengineering* 106 (4): 627–637.

Bruggen, B. Van der, and J. Geens. 2009. "Nanofiltration." In *Advanced Membrane Technology and Applications*. Hoboken, NJ: John Wiley & Sons.

Burnouf, T., and M. Radosevich. 2003. "Nanofiltration of plasma-derived biopharmaceutical products." *Haemophilia* 9 (1): 24–37. doi:10.1046/j.1365-2516.2003.00701.x.

Camacho, L. M., L. Dumée, J. Zhang, J.-de Li, M. Duke, J. Gomez, and S. Gray. 2013. "Advances in membrane distillation for water desalination and purification applications." *Water* 5 (1): 94–196.

Charcosset, C.. 2012a. "2—Ultrafiltration." In *Membrane Processes in Biotechnology and Pharmaceutics*, pp. 43–99. Amsterdam, the Netherlands: Elsevier.

Charcosset, C.. 2012b. "3 Microfiltration." In *Membrane Processes in Biotechnology and Pharmaceutics*, pp. 101–141. Amsterdam, the Netherlands: Elsevier.

Charcosset, C.. 2012c. "5—Membrane chromatography." In *Membrane Processes in Biotechnology and Pharmaceutics*, pp. 169–212. Amsterdam, the Netherlands: Elsevier.

Chenette, H. C. S., J. M. Welsh, and S. M. Husson. 2017. "Affinity membrane adsorbers for binding arginine-rich proteins." *Separation Science and Technology* 52 (2): 276–286. doi:10.1080/01496395.2016.1206934.

Cheryan, M. 1998. *Ultrafiltration and Microfiltration Handbook,* Boca Raton, FL: CRC Press.

Christy, C., G. Adams, R. Kuriyel, G. Bolton, and A. Seilly. 2002. "High-performance tangential flow filtration: A highly selective membrane separation process." *Desalination* 144 (1): 133–136. doi:10.1016/S0011-9164(02)00301-6.

Clarkson, A. I., P. Lefevre, and N. J. Titchener-Hooker. 1993. "A study of process interactions between cell disruption and debris clarification stages in the recovery of yeast intracellular products." *Biotechnology Progress* 9: 462–467.

Clever, M., F. Jordt, R. Knauf, N. Räbiger, M. Rüdebusch, and R. Hilker-Scheibel. 2000. "Process water production from river water by ultrafiltration and reverse osmosis." *Desalination* 131 (1): 325–336. doi:10.1016/S0011-9164(00)90031-6.

Curcio, E., X. Ji, A. M. Quazi, S. Barghi, G. D. Profio, E. Fontananova, T. Macleod, and E. Drioli. 2010. "Hybrid nanofiltration–membrane crystallization system for the treatment of sulfate wastes." *Journal of Membrane Science* 360 (1–2): 493–498.

Dichtelmüller, H. O., E. Flechsig, F. Sananes, M. Kretschmar, and C. J. Dougherty. 2012. "Effective virus inactivation and removal by steps of Biotest Pharmaceuticals IGIV production process." *Results in Immunology* 2: 19–24. doi:10.1016/j.rinim.2012.01.002.

DiLeo, A. J., A. E. Allegrezza Jr, and S. E. Builder. 1992. "High resolution removal of virus from protein solutions using a membrane of unique structure." *Bio/Technology* 10: 182. doi:10.1038/nbt0292-182.

Doelle, H. W., J. S. Rokem, and M. Berovic. 2009. *Biotechnology—Volume VI: Fundamentals in Biotechnology*. Eolss Publishers.

Falsanisi, D., L. Liberti, and M. Notarnicola. 2009. "Ultrafiltration (UF) pilot plant for municipal wastewater reuse in agriculture: Impact of the operation mode on process performance." *MDPI* 872–885.

Fan, J., J. Luo, and Y. Wan. 2017a. "Aquatic micro-pollutants removal with a biocatalytic membrane prepared by metal chelating affinity membrane chromatography." *Chemical Engineering Journal* 327: 1011–1020. doi:10.1016/j.cej.2017.06.172.

Fan, J., J. Luo, and Y. Wan. 2017b. "Membrane chromatography for fast enzyme purification, immobilization and catalysis: A renewable biocatalytic membrane." *Journal of Membrane Science* 538: 68–76. doi:10.1016/j.memsci.2017.05.053.

Ghosh, R., E. M. V. Hoek, and V. V. Tarabara. 2013. "Application of membranes in biotechnology." In *Encyclopedia of Membrane Science and Technology*. John Wiley & Sons.

Ghosh, R. 2002. "Protein separation using membrane chromatography: Opportunities and challenges." *Journal of Chromatography A* 952 (1): 13–27. doi:10.1016/S0021-9673(02)00057-2.

Gode, F. and E. Pehlivan. 2005. "Removal of Cr (VI) from aqueous solution by two Lewatit-anion exchange resins." *Journal of Hazardous Materials* 119 (1–3): 175–182.

Grandison, A. S. 1996. "Chapter 5—Microfiltration." In *Separation Processes in the Food and Biotechnology Industries*, pp. 141–153. Woodhead Publishing.

Hammond, S. M., S. Boettcher, A. A. Caudy, R. Kobayashi, and G. J. Hannon. 2001. "Argonaute2, a link between genetic and biochemical analyses of RNAi." *Science* 293 (5532): 1146–1150. doi:10.1126/science.1064023.

Han, B., R. Specht, S. R. Wickramasinghe, and J. O. Carlson. 2005. "Binding Aedes aegypti densonucleosis virus to ion exchange membranes." *Journal of Chromatography A* 1092 (1): 114–124. doi:10.1016/j.chroma.2005.06.089.

Hansen, R. 1988. "Better market milk, better cheese milk and better low-heat milk powder with Bactocatch treated milk." *North European Food and Dairy Journal* 1: 5–7.

He, X.-M., G.-T. Zhu, W. Lu, B.-F. Yuan, H. Wang, and Y.-Q. Feng. 2015. "Nickel(II)-immobilized sulfhydryl cotton fiber for selective binding and rapid separation of histidine-tagged proteins." *Journal of Chromatography A* 1405: 188–192. doi:10.1016/j.chroma.2015.05.040.

Himstedt, H. H., X. Qian, J. R. Weaver, and S. R. Wickramasinghe. 2013. "Responsive membranes for hydrophobic interaction chromatography." *Journal of Membrane Science* 447: 335–344. doi:10.1016/j.memsci.2013.07.020.

Hosseini, K. M. and S. Nasiri. 2015. "Preparation of factor VII concentrate using CNBr-activated Sepharose 4B immunoaffinity chromatography." *Methods* 15: 18.

Hostettmann, K., A. Marston, and M. Hostettmann. 1986. *Preparative Chromatography Techniques*, Springer.

Hung, Y. T., L. K. Wang, and N. K. Shammas. 2012. *Handbook of Environment and Waste Management: Air and Water Pollution Control*, World Scientific Publishing Company.

Ireland, T., H. Lutz, M. Slwak, and G. Bolton. 2006. "Viral filtration of plasma-derived human IgG." *Pharmaceutical Technology Europe* 18 (3): 43–49.

Jiang, J., X. Sun, Y. Li, C. Deng, and G. Duan. 2018. "Facile synthesis of Fe_3O_4@PDA core-shell microspheres functionalized with various metal ions: A systematic comparison of commonly-used metal ions for IMAC enrichment." *Talanta* 178: 600–607. doi:10.1016/j.talanta.2017.09.071.

Guu, Y. K., and R. R. Zall. 1992. Nanofiltration concentration effect on the efficacy of lactose crystallization, *Journal of Food Science* 57 (3): 735–739.

Kern, G., and M. Krishnan. 2006. "Virus removal by filtration: points to consider." *Biopharm International* 19 (10).

Kress-Rogers, E. 1996. *Handbook of Biosensors and Electronic Noses: Medicine, Food, and the Environment*, Boca Raton, FL: CRC Press.

Kumar, P., N. Sharma, R. Ranjan, S. Kumar, Z. F. Bhat, and D. K. Jeong. 2013. "Perspective of Membrane Technology in Dairy Industry: A Review." *Asian-Australasian Journal of Animal Sciences* 26 (9): 1347–1358. doi:10.5713/ajas.2013.13082.

Le, T. T., A. D. Cabaltica, and V. M. Bui. 2014. "Membrane separations in dairy processing." *Journal of Food Research and Technology* 2 (1): 1–14.

Li, Y., N. Butler, and A. L. Zydney. 2016. "Size-based separation of supercoiled plasmid DNA using ultrafiltration." *Journal of Colloid and Interface Science* 472: 195–201. doi:10.1016/j.jcis.2016.03.054.

Marques, B. F., D. J. Roush, and K. E. Göklen. 2009. "Virus filtration of high-concentration monoclonal antibody solutions." *Biotechnology Progress* 25 (2): 483–491.

McGregor, W. C. 1986. *Membrane Separations in Biotechnology*, Vol. 61. New York: Dekker.

McMurry, J. E. 2014. *Organic Chemistry with Biological Applications*, Stamford, CT: Cengage Learning.

Nanda, D., K.-L. Tung, C.-C. Hsiung, C.-J. Chuang, R.-C. Ruaan, Y.-C. Chiang, C.-S. Chen, and T.-H. Wu. 2008. "Effect of solution chemistry on water softening using charged nanofiltration membranes." *Desalination* 234 (1): 344–353. doi:10.1016/j.desal.2007.09.103.

Orr, V., L. Zhong, M. Moo-Young, and C. P. Chou. 2013. "Recent advances in bioprocessing application of membrane chromatography." *Biotechnology Advances* 31 (4): 450–465.

Pabby, A. K., S. S. H. Rizvi, and A. M. S. Requena. 2015. *Handbook of Membrane Separations: Chemical, Pharmaceutical, Food, and Biotechnological Applications*, CRC Press.

Roberts, M. W. H., C. M. Ongkudon, G. M. Forde, and M. K. Danquah. 2009. "Versatility of polymethacrylate monoliths for chromatographic purification of biomolecules." *Journal of Separation Science* 32 (15–16): 2485–2494. doi:10.1002/jssc.200900309.

Roy, I. and M. N. Gupta. 2006. "Bioaffinity immobilization." In *Immobilization of Enzymes and Cells*, edited by Jose M. Guisan, 107–116. Totowa, NJ: Humana Press.

Samhaber, W. M. and M. T. Nguyen. 2014. "Applicability and costs of nanofiltration in combination with photocatalysis for the treatment of dye house effluents." *Beilstein Journal of Nanotechnology* 5: 476–484. doi:10.3762/bjnano.5.55.

Shi, W., H. Cao, C. Song, H. Jiang, J. Wang, S. Jiang, J. Tu, and D. Ge. 2010. "Poly(pyrrole-3-carboxylic acid)-alumina composite membrane for affinity adsorption of bilirubin." *Journal of Membrane Science* 353 (1): 151–158. doi:10.1016/j.memsci.2010.02.048.

Shi, W., Y. Shen, H. Jiang, C. Song, Y. Ma, J. Mu, B. Yang, and D. Ge. 2010. "Lysine-attached anodic aluminum oxide (AAO)–silica affinity membrane for bilirubin removal." *Journal of Membrane Science* 349 (1): 333–340. doi:10.1016/j.memsci.2009.11.066.

Starbard, N. 2009. *Beverage Industry Microfiltration*, New York: John Wiley & Sons.

Teepakorn, C., K. Fiaty, and C. Charcosset. 2015. "Optimization of lactoferrin and bovine serum albumin separation using ion-exchange membrane chromatography." *Separation and Purification Technology* 151: 292–302. doi:10.1016/j.seppur.2015.07.046.

Titchener-Hooker, N. J., D. Gritsis, K. Mannweiler, R. Olbrich, S. A. M. Gardiner, N. M. Fish, and M. Hoare. 1991. "Integrated process design for producing and recovering proteins from inclusion bodies." *BioPharm* 4 (4): 34–38.

Tun, C. M. and A. M. Groth. 2011. "Sustainable integrated membrane contactor process for water reclamation, sodium sulfate salt and energy recovery from industrial effluent." *Desalination* 283: 187–192. doi:10.1016/j.desal.2011.03.054.

Tutunjian, R. S. 1985. "Scale-Up considerations for membrane processes." *Bio/Technology* 3: 615. doi:10.1038/nbt0785-615.

Uragami, T. 2017. *Membrane Shapes and Modules*, Chichester, UK: John Wiley & Sons.

Van der Bruggen, B., M. Mänttäri, and M. Nyström. 2008. "Drawbacks of applying nanofiltration and how to avoid them: A review." *Separation and Purification Technology* 63 (2): 251–263. doi:10.1016/j.seppur.2008.05.010.

Van Reis, R. and A. Zydney. 2007. "Bioprocess membrane technology." *Journal of Membrane Science* 297 (1): 16–50. doi:10.1016/j.memsci.2007.02.045.

Van Reis, R., S. Gadam, L. N. Frautschy, S. Orlando, E. M. Goodrich, S. Saksena, R. Kuriyel, C. M. Simpson, S. Pearl, and A. L. Zydney. 1997. "High performance tangential flow filtration." *Biotechnology and Bioengineering* 56 (1): 71–82.

Wei, Y., X. Huang, R. Liu, Y. Shen, and X. Geng. 2006. "Preparation of a monolithic column for weak cation exchange chromatography and its application in the separation of biopolymers." *Journal of Separation Science* 29 (1): 5–13.

Wu, W., Z. Cheng, M. Liu, X. Yang, and D. Qiu. 2014. "C3HC4-type RING finger protein NbZFP1 is involved in growth and fruit development in Nicotiana benthamiana." *PLoS One* 9 (6): e99352.

Zhang, H., H. Nie, D. Yu, C. Wu, Y. Zhang, C. J. Branford White, and L. Zhu. 2010. "Surface modification of electrospun polyacrylonitrile nanofiber towards developing an affinity membrane for bromelain adsorption." *Desalination* 256 (1): 141–147. doi:10.1016/j.desal.2010.01.026.

Zhu, A., W. Zhu, Z. Wu, and Y. Jing. 2003. "Recovery of clindamycin from fermentation wastewater with nanofiltration membranes." *Water Research* 37 (15): 3718–3732. doi:10.1016/S0043-1354(03)00250-1.

Zhu, J., J. Yang, and G. Sun. 2011. "Cibacron Blue F3GA functionalized poly(vinyl alcohol-co-ethylene) (PVA-co-PE) nanofibrous membranes as high efficient affinity adsorption materials." *Journal of Membrane Science* 385–386: 269–276. doi:10.1016/j.memsci.2011.10.001.

5 Reverse Osmosis

Kai Ling Yu, Sho Yin Chew, Shuk Yin Lu, Yoong Xin Pang, and Pau Loke Show

5.1 INTRODUCTION

Reverse osmosis (RO) is a membrane-based demineralization separation technique that separates ions and dissolved solids from a solution. RO relies on the force from applied pressure to force water from a solution to move through a semipermeable membrane in a direction opposite to the natural forward osmosis process (Kucera, 2010b). A membrane in an RO system serves as a selective barrier, allowing selected components to penetrate while rejecting undesired components. Water molecules can pass through freely over the membrane, so larger particles such as suspended solids and dissolved ions are retained on the other side of the membrane. The RO separation technique is generally applied to water-based units and systems that can produce comparatively pure water. The effluent stream entering the RO unit is separated into two different streams: the purified low concentration stream is known as permeate; the concentrated rejected stream is known as retentate (Kucera, 2000).

RO is commonly applied over applications such as solvent purifications or recovery of dissolved solids in solution. RO has high rejection ability toward components such as bacteria, microorganisms, sand, and protein that are commonly present in daily life, the output from RO is of high quality, which could directly meet the requirement of boiler feedwater and stream generators (Ahmad et al., 2013). Therefore, it is often coupled with or serves as an alternative for ion exchangers for purification of water. It can significantly reduce the regeneration frequency and waste production from a stand-alone ion-exchanger unit. Apart from the generation of high-purity water for applications in industries such as pharmaceuticals and microelectronics, RO is also used in processing alcoholic and fruit beverages, flavorings and additives, and various dairy products; treating wastewater; and recovering process materials used in the manufacturing industries (Lipnizki et al., 2012).

The osmosis phenomenon through a semipermeable membrane was first documented in 1748 (Cheryan, 1998). In the 1850s, the osmotic properties of ceramic membranes were studied. However, current development of this technology dates back to the 1940s when the osmotic properties of cellophane were investigated, specifically the existence of air film between two cellophane membranes where osmosis takes place through evaporation and condensation. The evaporative theory of osmosis was then replaced by diffusion and solution of solute in today's studies. In 1959, study on the desalination potential of cellulose acetate films and polymer films with hydrophilic properties was carried out (Reid and Breton, 1959a).

The osmosis and RO by pressuring a solution against a flat film have brought the development of the first asymmetric cellulose acetate membrane with significant commercial viability. The membrane was found to possess a 10 times better flux performance compared to other existing membrane materials (Baker, 2012). In the 1960s and 1970s, rapid advancement for commercialization of RO membranes led to the development of a solution-diffusion model that could very well represent empirical data and transport mechanism of membranes.

Since the 1970s, research has been focused on the development of new membranes that consist of properties such as higher throughput and better rejection of solutes at lower operating pressures and cost. Various efforts for surface altering have been carried out, and Hydranautic's membrane ESNA1-LF has been commercialized by Lenntech in municipal water treatment due to its effectiveness in removing organics and rejecting up to 91% of solid particles. Membrane feed pacer has also been investigated to minimize concentration polarization and pressure drop of the membrane (Kucera, 2011). Distance, angle, thickness, and orientation of the spacer were characterized. Studies show that lower cross-flow velocity has the most significant impact in terms of minimizing biofouling compared to other spacer modifications (Uemura and Henmi, 2008). With the advancement and development of nanotechnology, knowledge of nanomaterials has been applied to RO, such as nanocomposite membranes that allow the penetration of water with minimal applied pressure, and improvement on the membrane surface charges, roughness, and adhesion of microbes (Jeong et al., 2007; Kucera, 2010b).

This chapter describes the fundamentals of reverse osmosis (RO) technology, which are useful in understanding the transport mechanism and designing an RO system for different applications. This chapter then elaborates on the mathematical models established by various researchers. The models could assist in the evaluation of the impact of external factors in affecting membrane performance. Appropriate guidelines are provided in this chapter to minimize the particular elements that could disturb RO operations. Apart from mechanisms and theories, membrane materials that support RO separation are illustrated as a guideline for designing a more appropriate and efficient system for designated applications. Several examples of bioprocesses that apply RO technology in industry can be found in this chapter, together with the advantages and limitations of this particular system. This chapter concludes with a few new findings that could bring evolutionary improvement to RO technology.

5.2 MECHANISMS OF RO

5.2.1 Process Fundamental

The principles of how an RO system works is based on the fundamentals of a natural forward osmosis (FO) process. FO occurs when water travels through a semipermeable membrane driven by natural osmotic pressure from a dilute solution to a solution with a high concentration of dissolved solid. FO will take place continuously until the concentration of both sides of the membrane is equalized. Figure 5.1 illustrates the difference in water movement between FO and RO.

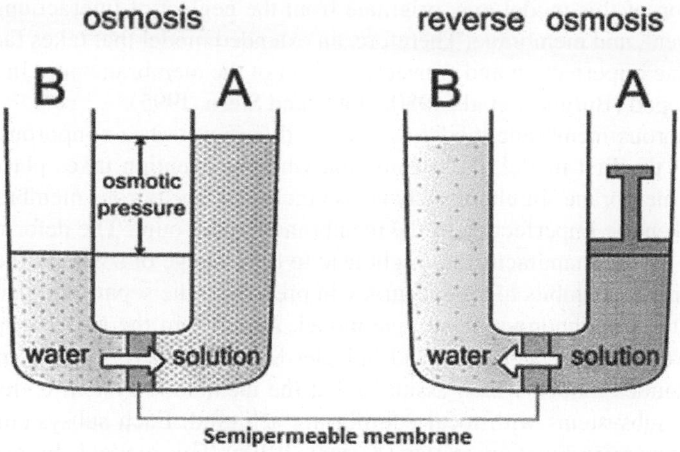

FIGURE 5.1 Schematic view of water movement in forward osmosis (FO) and reverse osmosis (RO).

The unit and equipment setup for RO is very similar to FO. As suggested by the word *reverse*, water flow in an RO system is opposite of its natural osmotic phenomenon. This is achievable by exerting a pressure that is greater than the osmotic pressure to the membrane side, which consists of a higher concentration of dissolved solids. Elevated pressure is required to overcome the resistance deployed by the membrane itself. For example, to purify brackish water containing 1500 ppm impurities, the pressure range for RO operation is from 150 to 400 psi (Kucera, 2010c). The applied pressure forces water to travel against the concentration gradient, from a concentrated solution to a relatively dilute solution. The dissolved solutes are retained on the other side of the membrane and the demineralized feed is then purified. Unlike FO, where the concentration of both sides of the membrane achieve equilibrium at the end of the process, RO produces a highly concentrated retentate and a comparatively pure permeate.

From the 1960s, when several transport theories of membrane were developed, researchers have also structured mathematical models to predict and study the membrane behavior and performance under different circumstances. The models were built with the assumption of the membrane at steady state or equilibrium condition. However, as the assumptions and the complexity of the theories differ, there are three categories of transport models in general. Lonsdale and others have proposed a nonporous membrane model: this solution-diffusion theory suggests that membranes are nonporous and the transport of material through the membrane is by dissolving into the membrane matrix and finally leaving the system by diffusion. Although the nonporous membrane model requires only two parameters to be characterized, researchers have pointed out that this model is only limited to membranes with low water content; especially for RO systems with organics, the water and solute flux could not be accurately described. Burghoff and coworkers suggested that the deviation of this model may originate from the neglect of interactions between solute, solvent, and membrane. Therefore, an extended model that takes factors such as membrane imperfection and convective effect of the membrane pore into account were developed (Burghoff et al., 1980; Noble and Stern, 1995).

In the porous membrane model, where the diffusion over a nonporous region is assumed as the first model, the assumption where convention takes place through the porous membrane simultaneously is also included. The porous membrane model has also taken the imperfection of the membrane into account. The deformity of the membrane during manufacturing might lead to the leakage of solute to the opposite side. This model exhibits higher accuracy in predicting the separation performance compared to the nonporous membrane model. Apart from the two transport models as above, Kadem-Katchalsky and Spiegler-Kedem model gives the irreversible thermodynamics model, which assumes that the membrane system is divided into ultra-small subsystems with local equilibrium achieved. Each subsystem then has its thermodynamics equation written (Kucera, 2010c). This model is highly complex and still in the academic development stage; hence, is the least applied transport model in the RO industry.

RO applications commonly use cross-flow filtration in the system, as shown in Figure 5.2. Cross-flow filtration is relatively similar to fluid flow. As the effluent flow direction is tangential to the membrane, turbulent flow occurs in the bulk solution, while no convective flow is observed at the surface of the boundary. In a membrane process where water passes through the membrane and rejected solutes deposit around the membrane surface, transport between surfaces of boundary to membrane is carried out by diffusion. However, as solutes tend to accumulate next to the membrane surface, the surrounding around the membrane is at a higher concentration compared to the incoming effluent stream. The concentration difference between the bulk and membrane surface is known as concentration polarization (Kucera, 2015).

Concentration polarization has a negative effect on the performance of an RO system. The membrane boundary layer has natural hydraulic resistivity to water flow across the membrane. Second, with a higher concentration of solutes deposit around the membrane boundary, the osmotic pressure increases, thus decreasing the driving force for water penetration. In order to minimize the effect of

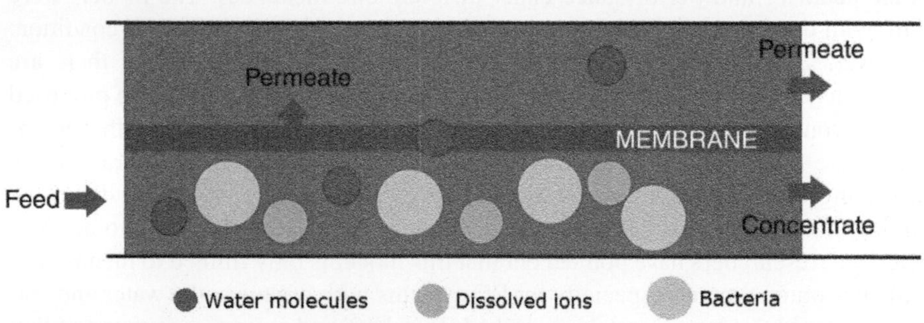

FIGURE 5.2 Schematic view of cross-flow filtration in reverse osmosis (RO) system.

concentration polarization, the membrane flux, which is the rate of water being forced across the membrane, should be minimized. At a high membrane flux, the effluent flow rate is high, bringing solutes to the membrane surface at a faster rate and thus increasing the rejection and accumulation rate of dissolved solids at the boundary. Due to the increasing concentration of solids in the feed stream, the membrane is found to be more vulnerable to fouling and scaling. Apart from adjusting the membrane flux, optimizing the bulk feed flow can also reduce the effect of concentration polarization. The flow across the membrane affects the boundary layer thickness. By decreasing the thickness of the boundary layer with a higher concentrate velocity, the effect of concentration polarization can be reduced (Kucera, 2000; Meyer et al., 2000).

5.2.2 System Design

The main factors that must be accounted for while designing an RO system consist of feed source quality, temperature, pH, pressure, flux, and flow rate. The quality of the inlet stream is important to design the size of the system and to select the particular type of pretreatment needed for RO prior to the separation. The ions present, alkalinity, total organics content, color, pH, and temperature are the basic analytics to be included for the RO inlet stream. The higher the temperature, the lower the driving force from pressure that is needed to achieve the same amount of membrane output. On the other hand, a lower temperature results in the decrease of salt passage, leading to an increase in permeate production and quality. Hence, it is advisable to operate an RO system at a lower temperature (Redondo and Lomax, 1997). As a function of temperature, pH affects the stability and rejection capability of the membranes. For an RO membrane, the rejection of a majority of species takes place at a pH value of 7–7.5. At lower or higher pH, the state of ions and molecular level of the membrane change. Therefore, the rejection of molecules is poorer at low and high pH values (Petelska and Figaszewski, 2000).

RO is a pressure-driven process, and the operating pressure in the system directly affects the driving force across the membrane. With higher operating pressure, the amount of water forced to the other side of the membrane increases, giving higher flux in the system. In theory, salt transport in RO is unaffected by pressure. However, as more permeate is produced at higher pressure and the concentration of salt in permeate is lower, it seems that the salt rejection of RO is increased at higher pressure (Ahmad et al., 2013). Continuous stable flow of designed inlet and recovery streams are preferable in an RO system. Fluctuating flow during separation could lead to the occurrence of concentration polarization and an overall low performance of the system (Porter, 1990). Membrane number and the design array of an RO system play a critical role in maximizing the cross-flow velocity of the membrane for recovery purposes. Increasing the stage number can provide better recovery: a single stage membrane system can achieve an estimated 50% recovery, but with a second stage installed in place, 75% of recovery can be achieved (Kucera, 2010a). An example of a two-stage RO system is presented in Figure 5.3.

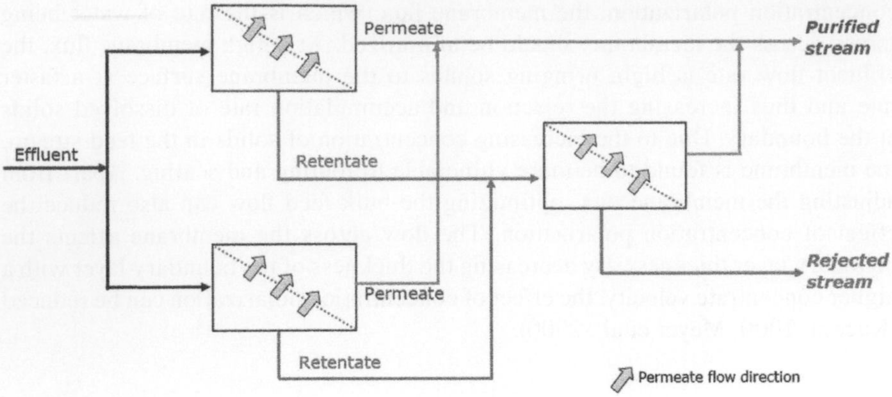

FIGURE 5.3 Overall view of two-stage reverse osmosis (RO) system.

5.3 RO MEMBRANES AND FOULING

5.3.1 RO Membranes and Materials

The performance of RO is affected by the properties of the membrane materials. In other words, the chemical structure and nature of the membrane determine the rejection and flux of the RO system. High rejection and flux, as well as high strength and robustness, is expected. However, a high-rejection membrane always comes with low flux, which leads to the desire to develop membranes that can achieve both high flux and high rejection. In the last few years, membranes that can balance between high flux and rejection have been developed (Lee et al., 2011). The most common RO membranes are cellulose acetate and polyamide. Cellulose acetate membranes have lower rejection rates and operate under higher pressure. To improve the drawbacks of cellulose acetate membranes, polyamide membranes were developed (Kucera, 2011). The various types of membranes are discussed below.

5.3.1.1 Cellulosic Membranes

Cellulose acetate (CA) membranes, also known as asymmetric membranes, were the first commercialize RO membranes developed by Leob and Sourirajan (1960). A CA membrane consists of a dense permselective skin and a thick supporting layer. Both layers have the same composition but different structure (Kucera, 2011). They are normally used to filter proteins, enzymes, biological fluids, and so on. CA membranes are simple to manufacture, tough, and chlorine resistance (Baker, 2004). They have chlorine toleration up to 1 ppm; therefore, they are suitable for high bacterial loading feed streams where chlorine sterilization is necessary (Baker, 2004). The smooth surface and nonpolar properties of CA membranes protect the membranes from being fouled (Kucera, 2011). However, CA membranes support the growth of microbial species: the membrane itself supplies the nutrient for the microbial growth. As a result, the microbial species metabolize the membrane and thus degrade it. These membranes are operated at 200–400 psig. Higher pressure

(>400 psig) causes membrane compaction, resulting in a much denser membrane with low water permeability (Kucera, 2011).

CA membranes are made by casting a film from a solution containing the cellulose acetate on a sturdy supporting layer (Kucera, 2011). The films are made of di- and triacetate, and combinations of the two (Kucera, 2011). The flux and rejection of CA membranes depend on the degree of acetylation of the polymer (Reid and Breton, 1959b; Rosenbaum et al., 1959), whereby the higher the acetyl content in the membranes, the higher the rejection. This means that the membranes containing high levels of acetyl have high selectivity. Unfortunately, the flux is decreased with an increase in the acetyl content, reflecting low flux at the high acetyl content (Lonsdale et al., 1965). Nevertheless, a small fraction of seawater desalination plants still use the cellulose triacetate (44.2 wt% of acetate) hollow fine fiber membranes because of their high percentage of rejection (99.6%) (Baker, 2004). However, commercial CA membranes usually contain 40 wt% of acetate with 98%–99% salt rejection and have an acceptable flux (Baker, 2004).

After the casting stage, the film is left to rest for 10–100 sec to evaporate the solvent to some extent (Kucera, 2011). The membrane is then immersed in a hot water bath for several minutes to produce a dense and more salt rejection permselective skin (Baker, 2004). The annealing process is used to modify the permselective skin of the CA membrane by eliminating the microspores of the membrane (Baker, 2004). In fact, the salt rejection and flux of CA membranes depend on the annealing temperature: the higher the annealing temperature, the higher the salt rejection, while the lower the flux. In summary, the final properties of CA membranes are determined by the annealing temperature (Sourirajan, 1970).

CA membranes are usually used below 35°C because higher temperatures may degrade the membranes, as well as further anneal them, leading to low flux (Kucera, 2011). The acetate group is susceptible to hydrolyzation under high acidic or alkaline conditions. Therefore, the pH of the feed is required to adjust to 4–6 to maintain the stability of the membranes (Vos et al., 1966).

5.3.1.2 Noncellulosic Membrane

Richter and Hoehn developed the first noncellulosic membrane, which later was commercialized by Du Pont with the name of B-9 Permasep® (Hoehn and Richter, 1980). B-9 Permasep® is an asymmetric, hollow-fiber membrane consisting of aromatic polyamide (PA) (Hoehn and Richter, 1980). Although its salt rejection and flux are comparatively low, it is more durable, stable, and versatile than the CA membrane (Beasley, 1977). In spite of the low water permeability, the hollow-fiber form of the membranes can overcome the low water flux problem and was used in seawater desalination plants until year 2000 (Richter and Hoehn, 1971).

Nevertheless, chlorine degradation was observed in the B-9 Permasep® membranes (Richter and Hoehn, 1971; Credali et al., 1978). Subsequently, a chlorine-resistant membrane consisting of polypiperazine-amides was developed to overcome this issue (Richter and Hoehn, 1971; Credali et al., 1978). The reduction of the amidic hydrogen in the membrane enhances the chlorine resistance; however, the salt rejection of the membrane is low (<95%), which reduced its commercial attractiveness (Parrini, 1983). Similarly, the sulfonated and carboxylated polysulfone are found to have a promising flux, yet the salt rejection was under the acceptable limit needed for commercialization (Brousse et al., 1976; Kurihara and Himeshima, 1991). In contrast, the polybenzimidazoline (PBIL) membranes show outstanding permselectivity even under adverse conditions, but they suffer from chlorine degradation and pressure compaction (Senoo et al., 1976).

5.3.1.3 Interfacial Composite Membranes

Interfacial composite (IFC) membranes, also known as thin-film composite membranes, are a single membrane fabricated by joining together two layers of different materials: a microporous support layer and a permselective skin. IFC membranes are made by impregnating the microporous polysulfone support with an aqueous amine solution. Next, the amine coating membrane is brought into contact with the acid chloride-containing hexane solution. A polymer thin film forms at the interface of the membrane as a result of the reaction between the impregnated membrane and the acid chloride-containing hexane solution (Cadotte 1981, 1985; Larson et al., 1981).

The surface of IFC membranes contains anion functional groups that attract the positive charge species to the membrane and foul the membrane. The reason for membrane fouling could be the rough surface structure of the membrane, where the foulants are trapped or held. Compared with the CA membrane, this membrane has a wider pH range (2–12) and can operate at a temperature up to 45°C. The operating pressure of IFC membranes (150–400 psig) is lower than that of CA membranes. This is because of the thin rejecting layer and the highly porous support layer of the IFC membranes, which decrease the resistance for permeate to pass through. The salt rejection of this membrane is usually higher than that of the CA membrane. For instance, using the 200 ppm sodium chlorine solution as a feed solution, this membrane can produces a salt rejection of about 99.7% at an operating pressure of 225 psi. However, IFC membranes have no resistance to chlorine compared to CA membranes (Kucera, 2011).

The first developed IFC membranes were known as NS-100 (nonpolysaccharides). NS-100 membranes were formed by the reaction between polythylenimine and toluene di-isocyanate (Porter, 1990). These membranes showed a significantly high salt rejection (>99%) and flux (30 L/m^2h) than CA membranes when tested with a sodium chloride solution of 3.5% at 100 bar operating pressure (Porter, 1990). NS-100 membranes can work at temperatures higher than 35°C, which is higher than the maximum allowable working temperature of the Leob-Sorirajan CA membranes (Porter, 1990). IFC membranes have superior low-molecular-weight dissolved organic solutes rejection. However, these membranes are extremely vulnerable to chlorine attack. Their permselective layer can be destroyed when exposed to even smaller ppb levels of chlorine. Also, the highly cross-linked configuration makes their surface very fragile (Porter, 1990). Table 5.1 shows the differences between CA and IFC membranes.

PA-300 (polyethylene amine) and RC-100 (regenerated cellulose) are IFC membranes formed by the polyamine reactant. Studies have found that PA-300 membranes have higher salt rejection (99.4%) and flux (1 m^3/m^2/day) at 70 bar when compared to the NS-100 membrane (Riley et al., 1976). Due to the reduction of amide functional groups in the structure, the PA-300/RC-100 membranes exhibit higher chlorine

TABLE 5.1
Comparison between CA and IFC Membranes in Terms of Characteristics, Advantages, and Disadvantages

Characteristics	Cellulose Acetate (CA) Membrane	Interfacial Composite Membrane
Type of membrane	Asymmetric membrane	Thin film composite asymmetric membrane
Percentage of salt rejection	Approximately 95%	Approximately 98%
Operating pH	4–6	2–12
Operating pressure	200–400 psig	145–400 psig
Operating temperature	≤35°C	≤45°C
Surface polarity	Non-polar	Anionic
Chlorine resistance	Up to 1 ppm	<0.02 ppm
Fouling resistance	Good	Fair
Surface coarseness	Smooth	Rough
Biological growth	Poor	Causes membrane fouling
Advantages		
	• Lower purchase cost • Resistant to chlorine in feed water up to a certain limit	• Excellent flux at lower pressure • A wide range of pH • High membrane stability with high resistance to hydrolysis • Resists compaction even at high pressure • A wide range of temperature • Superior salt rejection
Disadvantages		
	• Higher operating cost as high pressure is required for obtaining high flux • Narrow range for pH • Membrane hydrolyzes at high acidic and alkaline feed water • Compaction of membrane structure at high pressure • The narrow range of temperature • The growth of microbial resulting in membrane degradation	• Limited chlorine toleration. However, FT-30 membrane has better chlorine toleration than other IFC membranes.

Source: Kucera, J., *Reverse Osmosis: Design, Processes, and Applications for Engineers*, New York: John Wiley & Sons, 2011.

tolerance (Riley et al., 1976). Therefore, PA-300 in the form of spiral-wound modular was installed in Jeddah, Saudi Arabia, for desalination purposes (Hickman et al., 1979). The high biofouling resistance of RC-100 made it a successful choice for installation in the Umm Lujj II desalination plant (Light et al., 1988).

Membrane FT-30 (FilmTec) is a wholly aromatic polyamide membrane developed by Cadotte through the reaction between phenyleneditriamine and trimesoyl chloride (Cadotte, 1981). This membrane has unique surface features that can be described as ridges and valleys (Tarboush et al., 2008). The ridges and valleys can increase the water flux by increasing the effective surface area for water transport (Kwak et al., 1999). FT-30 produced 99.2% salt rejection with a water flux of 1 m^3/m^2/day at 55 bar operating pressure during seawater desalination tests (Cadotte, 1981). It can operate at a wide range of pH and has better resistance to chemicals, heat, and compression. Even though it does not thoroughly resist chlorine attack, FT-30 has the ability to withstand a chlorine attack up to 1000 ppm per hour. DOW FILMTEC® has successfully marketed this membrane in various products (Larson et al., 1981).

5.3.2 Membrane Fouling

Membrane fouling is a deposition of particles, molecules, salts, and so on, on the surface of a membrane and/or the inside of the pore walls of a membrane. This unfavorable phenomenon in RO systems causes membrane fouling that results in declination of permeate flux, lower product quality, higher energy demand, more frequent cleaning, and reduction in membrane lifecycle. Fouling can be divided into several categories, including colloidal fouling, biofouling, organic fouling, and inorganic scaling. Membrane fouling greatly depends on the feed water composition introduced to the RO system as well as feed water components' interactions with the membrane. In a production plant, more than one category of membrane fouling may occur. Therefore, fouling control is significant for preventing or minimizing fouling deposits.

5.3.2.1 Colloidal Fouling

Colloidal fouling, also known as silt fouling, is caused by deposition of fine suspended particles ranging from 1 to 1000 nm in size. Inorganic foulants and organic macromolecules are most commonly found in colloidal fouling. Inorganic foulant may consist of minerals; metal oxides/hydroxides; and organic macromolecules, including polysaccharides and proteins (Tang et al., 2011). The size, shape, and charge of the suspended particles influence the generation of colloidal fouling (Buffle et al., 1998). Interactions between foulant ions and membrane ions affect the fouling of membranes enormously. The colloidal aggregation rate is reflected by the collision frequency and coefficient of the attachment of particles. Declination of permeate flux and an increase in operation cost in RO due to the formation of a cake layer are drawbacks of colloidal fouling. The silt density index (SDI) can serve as a good predictor of the possibility that a particular feed water will generate colloidal fouling in RO. SDI measurement can be carried out by recording the time required for the desired feed water at fixed volume to flow through a microfiltration membrane with a pore size of 0.45 μm. Feed water that is susceptible to colloidal fouling has a longer filtration time due to deposition of suspended materials on the filter membrane. The SDI measurement indicates whether the RO system can run for several years (SDI of 1 or less) or several months (SDI of 3 or less), or will result in frequent fouling (SDI of 3–5). Any SDI value beyond 5 is unacceptable. The maximum tolerable SDI may vary for different designs of membrane module.

5.3.2.2 Biofouling

Biological fouling is the proliferation and adhesion of microorganisms on a membrane surface. It is also known as the formation of biofilm to the extent that it exceeds the acceptable limit. Biofilms can be formed either by bacteria or by extracellular polymeric substances (EPSs), which come from the metabolism of microorganism such as bacteria (Yu et al., 2016). Biofilm undergoes three stages of development: induction, logarithmical growth, and plateau (Flemming, 1997). Biofilm formation can also be divided into attachment, reproduction, and detachment of microorganisms. Piping systems that have low flux zones or dead zones are prone to bacteria attachment that forms biofilms. Other factors such as the interaction between the membrane and microorganisms (Kang et al., 2004), characteristics of the membrane surface (Nguyen et al., 2016), properties of microorganisms (Camesano and Logan, 1998) and operating conditions (Habimana et al., 2014) also affect the generation of biofilm. During the reproduction stage of microorganisms, stronger EPSs may form when there are sufficient nutrients for microorganism to live. As a result, the cleaning of biofilm process becomes more difficult (Ben-Dov et al., 2016). Formation of strong EPSs also protects the microorganisms from being destroyed (Belila et al., 2016). Finally, microorganisms detach from existing biofilms when there are insufficient nutrients for survival due to increase in population. These microorganisms will now migrate to new sites, and the formation of biofilm process repeats. At this stage, it will be extremely difficult to control the formation of biofilm.

5.3.2.3 Organic Fouling

Organic fouling is caused by attachment of organic matters such as oil or grease on the surface of a membrane. Organic fouling is mainly due to the behavior or organic components in oil or grease, which includes polysaccharides, lipids, and other organic matter such as organic acids and cell components. This fouling is commonly found in industrial plants that apply RO for effluent stream purification, especially in wastewater treatment. There are three important interactions that affect organic fouling: foulant-foulant interactions, foulant-surface interactions, and chemistry of feed water. One of the major drawbacks of organic fouling is that the combination of dissolved organic matters on the membrane forms a complex structure that is susceptible to declination of permeate flux in RO. This phenomenon is sustained without any further actions, and it will increase the energy consumption of RO system due to the higher energy required to overcome the resistance of the fouling layer on membranes, which prevents feed water from passing through (Naidu et al., 2014). The molecular weight of organic matter is another factor that affects the fouling of membranes. It was found that low- and medium-molecular-weight organic components are responsible for the beginning of fouling, while high-molecular-weight organic components that have more than 50,000 Da are responsible for most of the fouling (Lee et al., 2008). In addition, different organic components behave differently in terms of fouling, and their behavior is usually affected by different factors. For example, the behavior of bovine serum albumin (BSA) in relation to fouling is affected by foulant-deposited–foulant interaction, while the behavior of octanoic acid, a type of fatty acid, is influenced by both calcium ions and pH (Ang and Elimelech, 2008). Calcium carbonate, calcium sulphate, silica complexes, barium sulphate, strontium sulphate, and calcium fluoride are the common salts that have a proclivity to produce scale in RO.

5.3.2.4 Inorganic Scaling

Inorganic scaling is the precipitation of inorganic substances such as dissolved metal salt on the surface and/or the pores of membranes (Khayet, 2016). Scaling is the result of deposition of inorganic components on a membrane surface and/or the membrane pores when the inorganic components become supersaturated. Supersaturation of inorganic components occurs when the equilibrium solubility of the component has exceeded its limit. The components then undergo nucleation and homogenous or heterogeneous growth, which forms crystals that block the pathway of feed water to permeate through the membrane (Al-Amoudi and Lovitt, 2007). The characteristics of the membrane, the operating conditions of the RO system, and the composition of feed water have an impact on inorganic fouling. Scalants that deposit on the membrane surface and pores are difficult to clean. Backwashing may not be sufficient in removing scalants (Shirazi et al., 2010). A scaling inhibitor can be used to mitigate the deposition of inorganic components, especially calcium and magnesium. Chemicals such as polycrylate and polyphosphate can be used as antiscalants to minimize the precipitation of insoluble salts by maintaining the salt in the feed water solution even when the equilibrium solubility has exceeded its limit.

5.3.3 Membrane Fouling Control

Membrane fouling control is crucial and must be able to address all the problems in fouling effectively. Control strategies to be implemented should be first evaluated based on the category of fouling as well as the operating conditions of the RO system and the composition of fouling components present in the feed water to ensure that the fouling problem can be overcome. Membrane fouling can be controlled by implementing strategies such as pretreatment of feed water before it enters the RO system, surface modification of the membrane, and periodic membrane cleaning.

5.3.3.1 Pretreatment

Pretreatment of membranes is widely used in RO systems as a fouling control method to minimize the deposition of components on membranes. A variety of pretreatments is available for RO systems, and they vary from one and another depending on the category of fouling. When selecting a pretreatment method for a RO system, the composition of the components as well as their interaction with the membrane must be known. The list of pretreatment methods for RO includes flocculation, coagulation, additions of disinfectant, granular media filtration, addition of antiscalant, and many more. Flocculation and coagulation methods allow suspended particles in feed water to clump into large aggregates to ease the separation process. This method usually comes with sedimentation that separates the large aggregate from feed water. Thus, fouling caused by colloids can be minimized. On the other hand, addition of disinfectant is most commonly used to destroy microorganism in feed water. This method includes addition of chlorine, ultraviolet exposure, and ozone injection (Song et al., 2016). This pretreatment method is mostly used to control biofouling. Ultraviolet exposure and ozone injection are usually least preferred in comparison to chlorine because their operating costs are high. However, an additional step to remove chlorine before introducing feed water to the RO system must be done for the chlorine disinfection method because the presence of chlorine in feed water may result in degradation of the membrane. Although ozone injection is comparatively expensive compared to the addition of chlorine, the residual in ozone does not bring significant impact on RO system which eliminates the requirement of addition residual separation unit. Removal of suspended particles and microorganisms can also be done by granular media filtration, which consists of a porous granular medium. Activated carbon, sand, diatomaceous earth, and cotton are examples of granular material. As for organic scaling, addition of an antiscalant can minimize the scaling effect by altering the chemical and physical properties of ions present in feed water. However, selection of the antiscalant is crucial because the antiscalant must not cause any secondary fouling after removal of the target scalants. Ion exchange resins, commonly known as water softeners, can be an alternative for inorganic fouling control.

5.3.3.2 Surface Modification of the Membrane

The characteristics of the membrane surface, such as membrane morphology, hydrophobicity, charges, and chemical groups, play an important role in membrane fouling in RO systems (Saqib and Aljundi, 2016). Membranes with smooth and hydrophilic surfaces are often less likely to undergo membrane fouling compared to those membranes with rough and hydrophobic surfaces. Most of the time, a hydrophilic membrane is used when the corresponding foulant is hydrophobic because the foulant will be repelled by the membrane. However, if the foulant is hydrophilic, then a hydrophilic membrane is deemed unsuitable because it will attract the hydrophilic foulant and thus induce fouling of membrane. On the other hand, a membrane with a smooth surface is preferred over one with a rough surface because a rough membrane tends to accumulate the foulant easily. Modification of the membrane surface to minimize the fouling effect is most commonly used in the industry. Surface modification can be done by surface coating and surface grafting, which alter the physical and chemical properties of the membrane, respectively. Incorporation of nanoparticles in commercial polyamide thin-film composite RO membranes is one of the recent technologies for surface modification. For example, silver nanoparticles embedded on an RO membrane improved the membrane antimicrobial property, which helps in reducing biofouling. Carbon nanotube (CNT) incorporated membrane greatly reduces membrane fouling by enhancing the hydrophilic properties of the RO membrane.

5.3.3.3 Membrane Cleaning

Periodic membrane cleaning can significantly prolong the lifecycle of a membrane by minimizing the rate of fouling. The efficiency of cleaning can be evaluated by looking at the removal of resistance incurred by membrane fouling and the recovery of flux. A variety of membrane cleaning methods are available, including physical, chemical, biological, and enzymatic cleaning. All these cleaning methods aim to reduce the adhesion between the membrane surface and the foulant and thus allow the foulant to detach from the membrane at the end of cleaning process. For membrane cleaning using a chemical reagent, the reaction of the reagent with the fouling component and its compatibility with the membrane must be considered to ensure that fouling is reduced or eliminated without damaging the RO membrane. Chemical cleaning reagents such as acids are found to be effective for removal of scalants on membranes, whereas alkaline reagents are most commonly used for cleaning of inorganic foulants and biofoulants. Some of the common acids used in industry are nitric acid, sulfuric acid, and hydrochloric acid. Sodium hydroxide is the commonly used alkaline for treating biofouling and organic fouling. Besides acid and alkaline cleaning reagents, surfactants such as sodium dodecyl sulfate (SDS) can also be used as chemical reagents for treating colloidal fouling. The combination of physical and chemical membrane cleaning, where chemical cleaning is first applied to loosen the foulant layer and the physical cleaning then removes the foulant via fluid shear, can result in more efficient cleaning. However, this combination method is only considered when the RO system has a high fouling rate because it increases the overall cost of membrane cleaning.

5.4 APPLICATIONS

RO is most commonly found in both upstream and downstream in bioprocessing. Application of RO in upstream bioprocessing usually involves concentration of the desired biomaterial, such as concentrate conditioning for microalgae, separation and concentration of lactate from cassava fermentation broth, and concentrating flavoring for sweetener production. For downstream bioprocessing, RO is applied in the purification of water effluent, such as the removal of antibiotics from wastewater. The applications of RO in both upstream and downstream bioprocessing will be discussed in this section.

5.4.1 Applications in Upstream Processing

RO is a common purification method for waste disposal in industry. Water is purified using RO by concentrating the disposal in one of the chambers while water permeates through the membrane to produce RO water. The membrane-based, pressure-driven RO technology has been widely applied to various industries since its early development in the nineteenth century. Wastewater treatment processes lead in the use of RO, which includes desalination processes and sewage and effluent treatments. With further development in RO technology, the energy demand, process design, separation membrane materials, and overall cost have been improved.

Recent application of RO is in the upstream processing using RO concentrate for microalgae cultivation (Zhang et al., 2017). Wastewater contains essential nutrients such as nitrate and phosphate, and wastewater can be a substitute artificial medium by concentrating the nutrients using RO into a chamber during microalgae cultivation

(Joss et al., 2011). Although RO is an energy-intensive separation method, the high value of microalgae after cultivation can offset the cost of operation.

Separation and concentration of lactate from cassava fermentation broth by RO is also applicable in upstream processing. Production of polylactic acid derived from cassava root biomass consists of three stages: fermentation, separation-concentration, and polymerization. After fermentation, separation and concentration of lactic acid is required before polymerization of the lactic acid can take place to produce polylactic acid. RO is applied in the separation of lactic acid from fermentation broth because RO is relatively less costly compared to separation processes using distillation (Sekiguchi and Kokugan, 2010). Besides separation, the RO system also concentrates the lactic acid in one of the chambers in which the fermentation medium, unreacted cassava biomass, and strain are separated from lactic acid.

Advancements in RO have led to wider commercialization opportunities. The food industry has benefitted in its selective separation and concentrating processes for better purification such as concentrating food and vegetable juices, alcoholic beverages, as well as dairy products. By implying RO technology, concentration of sugar syrup can be carried out in a lower-energy demand process without affecting product quality. An example of RO concentration of sugar syrup from 13% to 30% by using commercialized polyamide PA300 or Filmtec FT30 RO membranes shows that the cost of such technology is relatively lower compared to evaporation (Madaeni et al., 2004). RO technology is simple and convenient to use in the processing of all categories of sugar syrup with various concentrations without affecting its color, composition, properties, and flavor. The collection of results suggests that this technology benefits the sugar processing industry by retaining the color, flavor, and properties of the syrup regardless of the sap concentration (Van Den Berg et al., 2005).

RO is a pressure-driven technology, and unlike conventional thermal processes, which have different energy demands for different types of sap, the amount of energy needed by RO to process various syrup is 33% (Wenten and Khoiruddin, 2016) lower than evaporators. Syrup production by RO increases the profitability of the production by reducing the time, energy, and complexity of the process without affecting the quality of the syrup produced. In spite of the high performance of RO membranes, this technology has limitations when it comes to sugar processing, including the requirement of pretreatment for enzymatic inactivation prior to the separation, the possibility of membrane fouling, and limits to the concentration of the final product caused by osmotic pressure. Hence, further development of RO technology in terms of membrane selectivity and fouling resistivity is required to improve the overall performance of this technique.

5.4.2 Applications in Downstream Processing

Other than upstream processing, the application of RO systems can be seen in downstream processing such as in the pharmaceutical industry. The increasing demand for pharmaceutical products in the past few years has raised the issue of pollution caused by the released of pharmaceutical residues to the environment as a result of the manufacturing processes and improper disposal. Pharmaceutical waste, such as antibiotics, steroids, hormones, and others, have been found in various environments

(Erickson, 2002). Among all the pharmaceutical wastes, antibiotics are one of the biggest concerns because they have the potential to cause resistance among the bacterial populations (Erickson, 2002). Antibiotics are persistence, bioaccumulative, and bioactive, and they can induce adverse effects to the environment even trace amounts (ng/L to low, μg/L) (Hernando et al., 2006). However, most of the wastewater treatment plants are not able to eliminate the highly polar pollutants such as antibiotics (Xu et al., 2007). As a result, they are transported to the surface water and reach the ground water. Eventually, they contaminate the drinking water treatment plants, which are not designed to remove the antibiotics.

Research on the effectiveness of RO for the removal of antibiotics was carried out by Gholami et al. (2012). A pilot scale RO skid was set up; it included a feed tank, heater, pump, and a membrane module that consisted of a spiral-wound thin-film polyamide membrane. Two different molecular weights of antibiotics, amoxicillin (365.4 g/mol) and ampicillin (349.91 g/mol), were used in the study. The results revealed that the salt rejection and flux were influenced by pressure, temperature, and pH. The rejection and permeate flux for amoxicillin was 99.36% and 18.5 $L/m^2/h$, respectively, whereas for ampicillin, it was 98.8% and 18.73 $L/m^2/h$, respectively, at the operating pressure of 13 bar, reflecting that the process was dominated by the size exclusion effect. Košutić, Dolar et al. (2007) carried out the experiments to test the rejection of antibiotics from a model pharmaceutical wastewater using RO. Two different types of membranes were used: XLE® and HR95PP®. The tested antibiotics included sulfadiazine, sulfamethazine, sulfaguanidine, oxytetracycline, trimethoprim, and enrofloxacin. The results showed that both membranes offered acceptably high rejection rates exceeding 98.5% for all the tested antibiotics. Similarly, the paper reported that the mechanism of rejection was the size exclusion effect.

RO has also been widely used in desalination of seawater for decades. Removal of boron in the form of boric acid in seawater is one of the applications. The presence of excessive boron in seawater causes serious health issues and ecological damage due to boron poisoning. Therefore, removal of boron in seawater is necessary to produce safe drinking water. RO was found to have considerably higher boron rejection than nanofiltration during standard operating conditions, which is low pH. Usually an RO system with two or more passes is commonly found for reducing the level of boron in seawater to desired levels. The number of passes is selected based on the conditions of feed water. To enhance the removal of boron, the first pass is normally operated at pH 6–7, whereas the second pass usually has slight elevation of pH up to 11. Another factor affecting boron rejection is the operating temperature. An increase in temperature was found to have better boron rejection due to a decrease in the concentration polarization of boric acid (Tu et al., 2010).

5.5 ADVANTAGES AND LIMITATIONS OF AN RO SYSTEM

RO is most commonly used to purify industrial effluent and comes with several advantages, such as reduced energy consumption, simple operation design, and environmental friendly. However, undeniable limitations require attention. The advantages and limitations of RO systems are listed in Table 5.2.

TABLE 5.2
Advantages and Limitations of Reverse Osmosis

Advantages/Limitations	Explanation	References
Less energy consumption	• Unlike other separation units like distillation, evaporation, and crystallization, RO is a separation and concentration process that does not require any phase transition. • As long as the osmotic pressure is low, the energy consumption of RO is relatively lower than a separation process that involves phase change.	Wenten and Khoiruddin (2016)
No solvent is required	• Does not require any solvent to remove unwanted components in water, which saves the cost of chemicals as well as a solvent recovery unit.	Wenten and Khoiruddin (2016)
Environmental friendly	• Separation and concentration are carried out by physical means. It does not consume or release any harmful chemicals to the environment.	Song et al. (2003)
Simple design	• Space-saving. • Easy to understand and only requires a short of time to pick up the working principle. • The modular design of the RO membrane enables the maintenance process to be carried out at any time without having to shut down the entire process plant.	Song et al. (2003)
Difficult to operate at large scale	• Small-scale RO unit: permeate flux is directly proportional to transmembrane pressure due to low osmotic pressure change along short channel of membrane module. • Large-scale RO system: the linear dependency of permeate flux toward transmembrane pressure deviates. As a result, higher transmembrane energy is required to maintain desired permeate flux. Subsequently, the energy consumption increases.	Song et al. (2003)
Requires pre-treatment before RO system	• The requirement of pre-treatment before RO to minimize membrane fouling implies additional unit is required to treat the feed water before introducing the feed water to RO. Consequently, the overall operation cost will be higher due to the addition of the treating process before RO.	Song et al. (2003)
Difficulties in in-situ membrane monitoring	• Lack of competency of in-situ and real-time membrane monitoring. • Membrane fouling can only be detected through a decrease in permeate flux after obvious fouling is formed. Therefore, the cleaning process will be more difficult as a thick layer of deposition is already formed. This increases the cost of maintenance because more chemicals or other cleaning reagents are required to clean the membrane thoroughly.	Jiang et al. (2017)

Membrane fouling is one of the biggest concerns in RO, and to date, few techniques have been developed to overcome the problem. Monitoring membrane fouling using ultrasonic time-domain reflectometry (UTDR) and electrical impedance spectroscopy (EIS) has been proposed (Fontananova et al., 2017; Sim et al., 2018). UTDR uses ultrasound waves to detect the thickness of a fouling layer by converting the return time and wave magnitude. EIS determines the fouling of a membrane by evaluating the changes in electrical properties of the membrane via electrodes installed with a resolution impedance spectroscope. These two techniques have given positive results in the detection of fouling and can resolve the problem. However, implementing those techniques is another cost for the overall RO process. Therefore, higher capital and operation costs of separation need to be solved for future implementation.

5.6 FUTURE PERSPECTIVES AND CHALLENGES

5.6.1 Carbon Nanotubes Membrane

Carbon nanotubes (CNTs) are carbon, tube-shaped objects having a diameter in the nanoscale (1–10 nm). The superior water transport efficiency of CNTs is promising in the purification and separation industries such as desalination and water purification. Studies have found that the transport properties of CNTs and biological membranes are comparable (Noy et al., 2007). CNTs are formed by the chemical vapor deposition (CVD) on the surface of the substrate coated with the metal catalyst. During the process, the heated substrate is exposed to the carbon-containing gas precursor in which the heat from the substrate decomposes the gas precursor and the carbon growth in the CNTs. An experimental analysis revealed that the flow velocity in the polystyrene film membrane embedded in CNTs (7 nm) was extraordinarily high: four to five orders of magnitude greater than the value obtained from the Haagen-Poiseuille equation (Majumder et al., 2005). A similar result was obtained in another study using a membrane with <2 nm diameter CNTs (Holt et al., 2006). The reason for flow enhancement was believed to be the result of the formation of a water molecules layer inside the CNTs which lead to frictionless water flow and thus increases the water permeability (Kotsalis et al., 2004). Another factor was due to the creation of the vapor layer between the surface of the CNTs and the bulk fluid, resulting in the slug flow manner that leads to high flow efficiency (Hummer et al., 2001).

The transport of ions through the CNTs (1–2 nm) has been investigated, and the results showed that the ion rejection was below the acceptable limit for a desalination process (Fornasiero et al., 2008; Holt et al., 2006). The researchers suggested that the mechanism of ion transport in the CNTs was dominated by the electrostatic exclusion rather than steric effects (Fornasiero et al., 2008; Holt et al., 2006). Therefore, Majumder and his group have demonstrated methods to change the selectivity of the CNT membrane to different ions, namely, voltage-based gate control and CNT tip functionalization. The ion rejection through these membranes was not tested; however, these studies presented the idea of changing the pore properties of the CNT membrane to improve electivity (Majumder et al., 2005). A study leading to a patent discovered that functionalized CNTs with octadecylamine could improve the solubility of the organic solvent. The results from the study stated that the membrane with CNTs (0.8 nm) showed higher salt rejection (97.69%) and water flux (44 L/m^2/day/bar) than the membrane without the CNTs (96.19% and 26 L/m^2/day/bar) (Ratto et al., 2010). However, the area of the membrane that was synthesized in the patent was just 47 mm. Thus, studies on the larger surface area membranes are necessary to commercialize the membrane embedded with CNTs.

5.6.2 Biomimetic RO Membranes

Aquaporins are the membrane proteins that are responsible for the transport of water between cells. The superb water transport properties of aquaporins has resulted in the study of membranes incorporating aquaporins (Agre, 2005). The studies have revealed that the membranes embedded with aquaporins have excellent water transport compared to RO membranes without the aquaporins (Kumar et al., 2007). Aquaporin-triblock polymer vesicles were used to study the water permeability. The results showed that there is at least an order of magnitude enhancement in the water flow compared to conventional thin film composite RO (Kumar et al., 2007). However, there is no result regarding the salt rejection. The salt rejection of the aquaporins is expected to be very high as the biological function of aquaporins is to allow water molecules to pass through (Gonzalez-Perez et al., 2009; Kumar et al., 2007; Taubert, 2007).

In 2005, a company named Aquaporins was established in Denmark to develop membranes incorporating aquaporins for industrial application. Recently, this company was awarded a patent for a method of developing aquaporin-embedded membranes (Jensen et al., 2010). The method suggested in the patent is the Langmuir Blodgett method, where the aquaporins are reconstituted into phospholipid bilayers. According to the patent, the aquaporins are embedded among the lipid bilayers and sandwiched between two layers of hydrophilic support. However, the patent does not report the data regarding water permeability and salt rejection of the membranes, while the concentration polarization and membrane fouling are stated.

5.6.3 Hybrid of Forward Osmosis-Electrodialysis-Reverse Osmosis

A single membrane separation process on its own is largely affected by the irreversible energy loss throughout the operation period. To overcome this limitation of membrane separation, various studies have targeted integrating membrane systems to increase the performance as well as the flexibility of the technology. Previous research has focused mainly on two-unit integrated systems, but recent findings analyzed systems with three units and found that the energy consumption was reasonable. A hybrid system of forward osmosis–electrodialysis–reverse osmosis (FO-ED-RO) has been studied to learn about the benefits of ED systems but reduce the energy demand of the overall system at the same time. In FO-ED-RO system, the FO unit is responsible for the main separation, while RO focuses on recovery. The ED unit is highly efficient in removing salt but its performance is limited by the concentration of the effluent. Dilute solution imposes high electrical resistance and affects the performance of ED membranes. Hence, the installation of pre- and post-treatment by FO and RO can effectively avoid the challenges faced by ED and enhances the entire separation system.

Osmosis is a pressure-driven process, and the FO unit creates high osmotic pressure in this hybrid system on the opposite side of a membrane with the addition of an ED unit. The output from the FO section is divided into concentrated and dilute streams. The two streams are directed into the ED unit, where the cationic and anionic exchange membranes draw the solutes to their respective regions by differentiating their particular charges. Ions of positive charge in the effluent travel through the cationic exchange membrane to the cathode, while the negative ions migrate to the anode via the anionic exchange membrane. As the ED membranes are charged, the separated ions are repulsed by the membrane charges and remain on its corresponding side. Concentrated stream from ED are sent back to FO as the draw solution to create continuous recycle streams in the system. The dilute stream of low concentration from ED enters the RO unit, where further recovery or purification of the effluent stream is carried out. Permeate of RO with lower concentration is the final product of this hybrid system, while the concentrated retentate is combined with ED concentrate stream to be recycled to FO unit. The investment cost and energy demand of the hybrid system are tightly related to factors such as current density, membrane resistivity, effluent concentration, and conductivity.

Numerous algorithm and calculation models have been developed to optimize the cost and performance of this hybrid system. If the limitations of each individual membrane system can be overcome by the hybrid unit with lower cost and reduced environmental effects, the development of such technology will be promising for wider applications of osmosis in various industries (Bitaw et al., 2016).

5.7 CONCLUSION

The early development, transport theories, and mechanisms of RO systems were described in this chapter. The common contributing factor for RO membrane failure was identified as fouling. Colloidal, biofouling, organic, and inorganic fouling are the categories of fouling affecting the performance of RO membranes. Consecutive control measures to prevent fouling were discussed, including conducting pretreatment for the effluent stream, modifying the surface membrane by coating, and altering the morphology of the surface membrane and clearing off the accumulated retentate regularly. The widely applied RO membranes for bioprocesses include cellulosic membranes, noncellulosic membranes, and interfacial composite membranes. Each membrane possesses different advantages and limitations, and further detailed studies on feed temperature, pH, membrane coating, and surface area will be needed to design a complete RO system for specific applications. RO technology is extensively applied in industries such as food and beverage, wastewater treatment, pharmaceuticals, and biotechnology. Despite limitations in the current technology, advancements new nanomaterials for fabrication of RO membranes and the incorporation of RO with other processing technologies such as forward osmosis and electrodialysis has enhanced the potential of RO for broader commercialization. The hybridization of technologies provides a new system with improved economic feasibility, environmental sustainability in terms of energy demand, and better performance in drawing solutes.

REFERENCES

Agre, P. 2005. "Membrane water transport and aquaporins: Looking back." *Biology of the Cell* 97 (6): 355–356. doi:10.1042/bc20050027.

Ahmad, F., K. K. Lau, A. M. Shariff, and Y. F. Yeong. 2013. "Temperature and pressure dependence of membrane permeance and its effect on process economics of hollow fiber gas separation system." *Journal of Membrane Science* 430: 44–55. doi:10.1016/j.memsci.2012.11.070.

Al-Amoudi, A. and R. W. Lovitt. 2007. "Fouling strategies and the cleaning system of NF membranes and factors affecting cleaning efficiency." *Journal of Membrane Science* 303 (1): 4–28. doi:10.1016/j.memsci.2007.06.002.

Ang, W. S. and M. Elimelech. 2008. "Fatty acid fouling of reverse osmosis membranes: Implications for wastewater reclamation." *Water Research* 42 (16): 4393–4403. doi:10.1016/j.watres.2008.07.032.

Baker, R. W. 2004. *Membrane Technology and Applications*. Hoboken, NJ: Wiley-Blackwell.

Baker, R. W. 2012. *Membrane Technology and Applications*. Hoboken, NJ: Wiley-Blackwell.

Beasley, J. K. 1977. "The evaluation and selection of polymeric materials for reverse osmosis membranes." *Desalination* 22 (1): 181–189. doi:10.1016/S0011-9164(00)88374-5.

Belila, A., J. El-Chakhtoura, N. Otaibi, G. Muyzer, G. Gonzalez-Gil, P. E. Saikaly, M. C. M. van Loosdrecht, and J. S. Vrouwenvelder. 2016. "Bacterial community structure and variation in a full-scale seawater desalination plant for drinking water production." *Water Research* 94: 62–72. doi:10.1016/j.watres.2016.02.039.

Ben-Dov, E., E. Ben-David, R. Messalem, M. Herzberg, and A. Kushmaro. 2016. "Biofilm formation on RO membranes: The impact of seawater pretreatment." *Desalination and Water Treatment* 57 (11): 4741–4748. doi:10.1080/19443994.2014.998294.

Bitaw, T. N., K. Park, and D. R. Yang. 2016. "Optimization on a new hybrid Forward osmosis-Electrodialysis-Reverse osmosis seawater desalination process." *Desalination* 398: 265–281. doi:10.1016/j.desal.2016.07.032.

Brousse, C. L., R. Chapurlat, and J. P. Quentin. 1976. "New membranes for reverse osmosis I. Characteristics of the base polymer: Sulphonated polysulphones." *Desalination* 18 (2): 137–153. doi:10.1016/S0011-9164(00)84098-9.

Buffle, J., K. J. Wilkinson, S. Stoll, M. Filella, and J. Zhang. 1998. "A generalized description of aquatic colloidal interactions: The three-colloidal component approach." *Environmental Science & Technology* 32 (19): 2887–2899. doi:10.1021/es980217h.

Burghoff, H. G., K. L. Lee, and W. Pusch. 1980. "Characterization of transport across cellulose acetate membranes in the presence of strong solute–membrane interactions." *Journal of Applied Polymer Science* 25 (3): 323–347. doi:10.1002/app.1980.070250301.

Cadotte, J. E. 1981. Interfacially synthesized reverse osmosis membrane. U.S. Patent No. 4,277,344.7.

Cadotte, J. E. 1985. "Evolution of composite reverse osmosis membranes." In *Materials Science of Synthetic Membranes*, 273–294. Washington, DC: American Chemical Society.

Camesano, T. A. and B. E. Logan. 1998. "Influence of fluid velocity and cell concentration on the transport of motile and nonmotile bacteria in porous media." *Environmental Science & Technology* 32 (11): 1699–1708. doi:10.1021/es970996m.

Cheryan, M. 1998. *Ultrafiltration and Microfiltration Handbook*. Boca Raton, FL: CRC Press.

Credali, L., G. Baruzzi, and V. Guidotti. 1978. Reverse osmosis anisotropic membranes based on polypiperazine amides. U.S. Patent No. 4,129,559.

Erickson, B. E. 2002. *Analyzing the Ignored Environmental Contaminants*. Washington, DC: ACS Publications.

Flemming, H. C. 1997. "Reverse osmosis membrane biofouling." *Experimental Thermal and Fluid Science* 14 (4): 382–391. doi:10.1016/S0894-1777(96)00140-9.

Fontananova, E., G. D. Profio, L. Giorno, and E. Drioli. 2017. "2.13 Membranes and interfaces characterization by impedance spectroscopy." In *Comprehensive Membrane Science and Engineering (Second Edition)*, Edited by Enrico Drioli, Lidietta Giorno and Enrica Fontananova, 393–410. Oxford, UK: Elsevier.

Fornasiero, F., H. G. Park, J. K. Holt, M. Stadermann, C. P. Grigoropoulos, A. Noy, and O. Bakajin. 2008. "Ion exclusion by sub-2-nm carbon nanotube pores." *Proceedings of the National Academy of Sciences of the United States of America* 105 (45): 17250–17255. doi:10.1073/pnas.0710437105.

Gholami, M., R. Mirzaei, R. R. Kalantary, A. Sabzali, and F. Gatei. 2012. "Performance evaluation of reverse osmosis technology for selected antibiotics removal from synthetic pharmaceutical wastewater." *Iranian Journal of Environmental Health Science & Engineering* 9 (1): 19–19. doi:10.1186/1735-2746-9-19.

Gonzalez-Perez, A., K. B. Stibius, T. Vissing, C. H. Nielsen, and O. G. Mouritsen. 2009. "Biomimetic triblock copolymer membrane arrays: A stable template for functional membrane proteins." *Langmuir* 25 (18): 10447–10450. doi:10.1021/la902417m.

Habimana, O., A. J. C. Semião, and E. Casey. 2014. "The role of cell-surface interactions in bacterial initial adhesion and consequent biofilm formation on nanofiltration/reverse osmosis membranes." *Journal of Membrane Science* 454: 82–96. doi:10.1016/j.memsci.2013.11.043.

Hernando, M. D., M. Mezcua, A. R. Fernández-Alba, and D. Barceló. 2006. "Environmental risk assessment of pharmaceutical residues in wastewater effluents, surface waters and sediments." *Talanta* 69 (2): 334–342. doi:10.1016/j.talanta.2005.09.037.

Hickman, C. E., I. Jamjoom, A. B. Riedinger, and R. E. Seaton. 1979. "Jeddah seawater reverse osmosis installation." *Desalination* 30 (1): 259–281. doi:10.1016/S0011-9164(00)88453-2.

Hoehn, H. H. and J. W. Richter 1980. Aromatic polyimide, polyester and polyamide separation membranes.

Holt, J. K., H. G. Park, Y. Wang, M. Stadermann, A. B. Artyukhin, C. P. Grigoropoulos, A. Noy, and O. Bakajin. 2006. "Fast mass transport through sub-2-nanometer carbon nanotubes." *Science* 312 (5776): 1034–1037. doi:10.1126/science.1126298.

Hummer, G., J. C. Rasaiah, and J. P. Noworyta. 2001. "Water conduction through the hydrophobic channel of a carbon nanotube." *Nature* 414 (6860): 188–190. doi:10.1038/35102535.

Jensen, P. H., D. Keller, and C. H. Nielsen. 2010. *Membrane for Filtering of Water*. Alexandria, VA: The United States Patent and Trademark Office.

Jeong, B. H., E. M. V. Hoek, Y. Yan, A. Subramani, X. Huang, G. Hurwitz, A. K. Ghosh, and A. Jawor. 2007. "Interfacial polymerization of thin film nanocomposites: A new concept for reverse osmosis membranes." *Journal of Membrane Science* 294 (1–2): 1–7.

Jiang, S., Y. Li, and B. P. Ladewig. 2017. "A review of reverse osmosis membrane fouling and control strategies." *Science of the Total Environment* 595: 567–583. doi:10.1016/j.scitotenv.2017.03.235.

Joss, A., C. Baenninger, P. Foa, S. Koepke, M. Krauss, C. S. McArdell, K. Rottermann, Y. Wei, A. Zapata, and H. Siegrist. 2011. "Water reuse: >90% water yield in MBR/RO through concentrate recycling and CO_2 addition as scaling control." *Water Research* 45 (18): 6141–6151. doi:10.1016/j.watres.2011.09.011.

Kang, S. T., A. Subramani, E. M. V. Hoek, M. A. Deshusses, and M. R. Matsumoto. 2004. "Direct observation of biofouling in cross-flow microfiltration: Mechanisms of deposition and release." *Journal of Membrane Science* 244 (1): 151–165. doi:10.1016/j.memsci.2004.07.011.

Khayet, M. 2016. "Fouling and scaling in desalination." *Desalination* 393:1. doi:10.1016/j.desal.2016.05.005.

Košutić, K., D. Dolar, D. Ašperger, and B. Kunst. 2007. "Removal of antibiotics from a model wastewater by RO/NF membranes." *Separation and Purification Technology* 53 (3): 244–249. doi:10.1016/j.seppur.2006.07.015.

Kotsalis, E. M., J. H. Walther, and P. Koumoutsakos. 2004. "Multiphase water flow inside carbon nanotubes." *International Journal of Multiphase Flow* 30 (7): 995–1010. doi:10.1016/j.ijmultiphaseflow.2004.03.009.

Kucera, J. 2011. *Reverse Osmosis: Design, Processes, and Applications for Engineers*. New York: John Wiley & Sons.

Kucera, J. 2000. "Reverse osmosis." In *Kirk-Othmer Encyclopedia of Chemical Technology*. New York: John Wiley & Sons.

Kucera, J. 2010a. "Basic flow patterns." In *Reverse Osmosis*, 85–93. Hoboken, NJ: John Wiley & Sons.

Kucera, J. 2010b. "Introduction and history of development." In *Reverse Osmosis*, 1–13. Hoboken, NJ: John Wiley & Sons.

Kucera, J. 2010c. "Reverse osmosis principles." In *Reverse Osmosis*, 15–19. Hoboken, NJ: John Wiley & Sons.

Kucera, J. 2015. "Reverse Osmosis Principles." In *Reverse Osmosis*, 19–24. Hoboken, NJ: John Wiley & Sons.

Kumar, M., M. Grzelakowski, J. Zilles, M. Clark, and W. Meier. 2007. "Highly permeable polymeric membranes based on the incorporation of the functional water channel protein Aquaporin Z." *Proceedings of the National Academy of Sciences of the United States of America* 104 (52): 20719–20724. doi:10.1073/pnas.0708762104.

Kurihara, M., and Y. Himeshima. 1991. "The major developments of the evolving reverse osmosis membranes and ultrafiltration membranes." *Polymer Journal* 23:513. doi:10.1295/polymj.23.513.

Kwak, S. Y., S. G. Jung, Y. S. Yoon, and D. W. Ihm. 1999. Details of surface features in aromatic polyamide reverse osmosis membranes characterized by scanning electron and atomic force microscopy. *Journal of Polymer Science Part B: Polymer Physics* 37 (13): 1429–1440.

Larson, R. E., J. E. Cadotte, and R. J. Petersen. 1981. "The FT-30 seawater reverse osmosis membrane—element test results." *Desalination* 38: 473–483. doi:10.1016/S0011-9164(00)86092-0.

Lee, E. K., V. Chen, and A. G. Fane. 2008. "Natural organic matter (NOM) fouling in low pressure membrane filtration—Effect of membranes and operation modes." *Desalination* 218 (1): 257–270. doi:10.1016/j.desal.2007.02.021.

Lee, K. P., T. C. Arnot, and D. Mattia. 2011. "A review of reverse osmosis membrane materials for desalination—Development to date and future potential." *Journal of Membrane Science* 370 (1): 1–22. doi:10.1016/j.memsci.2010.12.036.

Light, W. G., J. L. Perlman, A. B. Riedinger, and D. F. Needham. 1988. "Desalination of non-chlorinated surface seawater using TFCR membrane elements." *Desalination* 70 (1): 47–64. doi:10.1016/0011-9164(88)85043-4.

Lipnizki, J., B. Adams, M. Okazaki, and A. Sharpe. 2012. "Water treatment: Combining reverse osmosis and ion exchange." *Filtration + Separation* 49 (5): 30–33. doi:10.1016/S0015-1882(12)70245-8.

Loeb, S. and S. Sourirajan. 1960. *Sea Water Demineralization by Means of an Osmotic Membrane*. Los Angeles, CA: University of California, Department of Engineering.

Lonsdale, H. K., U. Merten, and R. L. Riley. 1965. Transport properties of cellulose acetate osmotic membranes. *Journal of Applied Polymer Science*. 9 (4): 1341–1362.

Madaeni, S. S., K. Tahmasebi, and S. H. Kerendi. 2004. "Sugar syrup concentration using reverse osmosis membranes." *Engineering in Life Sciences* 4 (2): 187–190. doi:10.1002/elsc.200401801.

Majumder, M., N. Chopra, R. Andrews, and B. J. Hinds. 2005. "Nanoscale hydrodynamics: Enhanced flow in carbon nanotubes." *Nature* 438 (7064):44. doi:10.1038/43844a.

Majumder, M., N. Chopra, and B. J. Hinds. 2005. "Effect of tip functionalization on transport through vertically oriented carbon nanotube membranes." *Journal of the American Chemical Society* 127 (25): 9062–9070. doi:10.1021/ja043013b.

Meyer, D. E., M. Williams, and D. Bhattacharyya. 2000. "Reverse osmosis." In *Kirk-Othmer Encyclopedia of Chemical Technology*. New York: John Wiley & Sons.

Naidu, G., S. Jeong, S. J. Kim, I. S. Kim, and S. Vigneswaran. 2014. "Organic fouling behavior in direct contact membrane distillation." *Desalination* 347: 230–239. doi:10.1016/j.desal.2014.05.045.

Nguyen, V., E. Karunakaran, G. Collins, and C. A. Biggs. 2016. "Physicochemical analysis of initial adhesion and biofilm formation of Methanosarcina barkeri on polymer support material." *Colloids and Surfaces B: Biointerfaces* 143: 518–525. doi:10.1016/j.colsurfb.2016.03.042.

Noble, R. D. and S. A. Stern. 1995. *Membrane Separations Technology: Principles and Applications*. Amsterdam, the Netherlands: Elsevier Science.

Noy, A., H. G. Park, F. Fornasiero, J. K. Holt, C. P. Grigoropoulos, and O. Bakajin. 2007. "Nanofluidics in carbon nanotubes." *Nano Today* 2 (6): 22–29. doi:10.1016/S1748-0132(07)70170-6.

Parrini, P. 1983. "Polypiperazinamides: New polymers useful for membrane processes." *Desalination* 48 (1): 67–78. doi:10.1016/0011-9164(83)80006-X.

Petelska, A. D. and Z. A. Figaszewski. 2000. "Effect of pH on the interfacial tension of lipid bilayer membrane." *Biophysical Journal* 78 (2): 812–817. doi:10.1016/S0006-3495(00)76638-0.

Porter, M. C. 1990. *Handbook of Industrial Membrane Technology*. New York: Noyes Publications.

Ratto, T. W, J. K. Holt, and A. W. Szmodis. 2010. Membranes with embedded nanotubes for selective permeability. U.S. Patent No. 7,993,524.

Redondo, J. A. and I. Lomax. 1997. "Experiences with the pretreatment of raw water with high fouling potential for reverse osmosis plant using FILMTEC membranes." *Desalination* 110 (1): 167–182. doi:10.1016/S0011-9164(97)81590-1.

Reid, C. E. and E. J. Breton. 1959a. "Water and ion flow across cellulosic membranes." *Journal of Applied Polymer Science* 1 (2): 133–143. doi:10.1002/app.1959.070010202.

Richter, J. W, and H. H. Hoehn. 1971. Selective aromatic nitrogen-containing polymeric membranes. U.S. Patent No. 3,567,632.

Riley, R. L., R. L. Fox, C. R. Lyons, C. E. Milstead, M. W. Seroy, and M. Tagami. 1976. "Spiral-wound poly (ether/amide) thin-film composite membrane systems." *Desalination* 19 (1): 113–126. doi:10.1016/S0011-9164(00)88022-4.

Rosenbaum, S., H. I. Mahon, and O. Cotton. 1959. Permeation of water and sodium chloride through cellulose acetate. *Journal of Applied Polymer Science*. 11 (10): 2041–2065.

Saqib, J. and I. H. Aljundi. 2016. "Membrane fouling and modification using surface treatment and layer-by-layer assembly of polyelectrolytes: State-of-the-art review." *Journal of Water Process Engineering* 11: 68–87. doi:10.1016/j.jwpe.2016.03.009.

Sekiguchi, M. and T. Kokugan. 2010. "Separation and concentration of lactate from cassava fermentation broth by reverse osmosis." *Membrane* 35 (5): 248–256.

Senoo, M., S. Hara, and S. Ozawa. 1976. Permselective Polymeric Membrane Prepared from Polybenzimidazoles. U.S. Patent No. 3,951,920.

Shirazi, S., C. J. Lin, and D. Chen. 2010. "Inorganic fouling of pressure-driven membrane processes—A critical review." *Desalination* 250 (1): 236–248. doi:10.1016/j.desal.2009.02.056.

Sim, L. N., T. H. Chong, A. H. Taheri, S. T. V. Sim, L. Lai, W. B. Krantz, and A. G. Fane. 2018. "A review of fouling indices and monitoring techniques for reverse osmosis." *Desalination* 434: 169–188. doi:10.1016/j.desal.2017.12.009.

Song, K., M. Mohseni, and F. Taghipour. 2016. "Application of ultraviolet light-emitting diodes (UV-LEDs) for water disinfection: A review." *Water Research* 94: 341–349. doi:10.1016/j.watres.2016.03.003.

Song, L., J. Y. Hu, S. L. Ong, W. J. Ng, M. Elimelech, and M. Wilf. 2003. "Performance limitation of the full-scale reverse osmosis process." *Journal of Membrane Science* 214 (2): 239–244. doi:10.1016/S0376-7388(02)00551-3.

Sourirajan, S. 1970. *Reverse Osmosis*. New York: Academic Press.

Tang, C. Y., T. H. Chong, and A. G. Fane. 2011. "Colloidal interactions and fouling of NF and RO membranes: A review." *Advances in Colloid and Interface Science* 164 (1): 126–143. doi:10.1016/j.cis.2010.10.007.

Tarboush, B. J. A., D. Rana, T. Matsuura, H. A. Arafat, and R. M. Narbaitz. 2008. "Preparation of thin-film-composite polyamide membranes for desalination using novel hydrophilic surface modifying macromolecules." *Journal of Membrane Science* 325 (1): 166–175. doi:10.1016/j.memsci.2008.07.037.

Taubert, A. 2007. "Controlling water transport through artificial polymer/protein hybrid membranes." *Proceedings of the National Academy of Sciences of the United States of America* 104 (52): 20643–20644. doi:10.1073/pnas.0710864105.

Tu, K. L., L. D. Nghiem, and A. R. Chivas. 2010. "Boron removal by reverse osmosis membranes in seawater desalination applications." *Separation and Purification Technology* 75 (2): 87–101. doi:10.1016/j.seppur.2010.07.021.

Uemura, T., and M. Henmi. 2008. "Thin-film composite membranes for reverse osmosis." In *Advanced Membrane Technology and Applications*, 1–19. New York: John Wiley & Sons.

Van Den Berg, A., T. Perkins, M. Isselhardt, M. A. Godshall, and S. Lloyd. 2015. "Effects of sap concentration with reverse osmosis on syrup composition and flavor." Maple Dig 54: 11–33.

Vos, K. D., F. O. Jr. Burris, and R. L. Riley. 1966. Kinetic study of the hydrolysis of cellulose acetate in the pH range of 2–10. *Journal of Applied Polymer Science* 10 (5): 825–832.

Wenten, I. G., and Khoiruddin. 2016. "Reverse osmosis applications: Prospect and challenges." *Desalination* 391: 112–125. doi:10.1016/j.desal.2015.12.011.

Xu, W. H., G. Zhang, S. C. Zou, X. D. Li, and Y. C. Liu. 2007. "Determination of selected antibiotics in the victoria harbour and the pearl river, south china using high-performance liquid chromatography-electrospray ionization tandem mass spectrometry." *Environmental Pollution* 145 (3): 672–679. doi:10.1016/j.envpol.2006.05.038.

Yu, W., Y. Yang, and N. Graham. 2016. "Evaluation of ferrate as a coagulant aid/oxidant pretreatment for mitigating submerged ultrafiltration membrane fouling in drinking water treatment." *Chemical Engineering Journal* 298: 234–242. doi:10.1016/j.cej.2016.03.080.

Zhang, D., K. Y. Fung, and K. M. Ng. 2017. "Reverse osmosis concentrate conditioning for microalgae cultivation and a generalized workflow." *Biomass and Bioenergy* 96: 59–68. doi:10.1016/j.biombioe.2016.11.004.

6 Chromatography

Kirupa Sankar Muthuvelu and
Senthil Kumar Arumugasamy

6.1 INTRODUCTION

Molecules are produced through a series of reactions that take place in a biological system or in synthetic chemistry. These molecules are recovered in addition to minute quantities of unwanted compounds referred as impurities. The target molecules must be isolated from the impurities for further applications (Pavia, 2005). This can be done on the basis of distinct physical properties such as molecular weight, boiling point, freezing point, crystallization, solubility, density, and chemical properties such as functional group and reactivity.

Chromatography consists a group of analytical protocols used for separating mixtures into their individual components (Walls et al., 2011). Before the identification of an unknown compound, it must be resolved to its constituents using a separation method. Since its invention, chromatography has been an important tool for qualitative and quantitative estimation of components present in a mixture.

6.1.1 History of Chromatography

The development of chromatography started in the mid-1800s with the work of a German dye chemist, F. F. Runge. Filter paper and water were used to separate the dye components (Ettre and Sakodynskii, 1993). Two reasons are responsible for the separation: the binding capacity of dye components toward the filter paper and also variations in the molecular weight among the dyes. In the 1860s, C. F. Schonbein and his student studied the rate of migration of various substances through filter paper (Ettre, 2000). Schonbein predicted that the separation was due to capillary action and named the technique capillary analysis. In 1906, a Russian botanist, Mikhail Tswett segregated six pigments present in plants, for example, chlorophyll (green) and carotenoids (yellow and orange), through a column loaded with calcium carbonate ($CaCO_3$) using petroleum ether as a solvent (Telepchak et al., 2004). Initially, chromatography was to be used only to separate the color components, which is the reason behind its name. The term *chromatography* was coined from the Greek words *Khroma*, for "color," and graphian, for "to write." Thus, it is described as identifying the constituents of a mixture based on color. In early 1949, Martin and

Anthony T. James developed gas chromatography and separated various natural compounds. Their work earned them the Nobel Prize in chemistry (Martin, 1950). Both Martin and James went on to further developments in gas chromatography. In 1954, N. H. Ray conducted much research on detection methods for analyzing the chromatography outputs (Ray, 2007). In 1987, Meir Wilchek and Pedro Cuatrecasas received the Wolf Prize in Medicine for the invention of affinity chromatography and its advanced applications in the field of biomedical sciences (Cuatrecasas et al., 1968; Horváth, 1988).

6.1.2 Branches of Chromatography

Chromatography can be classified as analytical or preparative. The goal of preparative chromatography is to sort out the constituents of a mixture for further use, and thus it is categorized under purification (Ettre and Hinshaw, 2008). The goal of analytical chromatography is to detect or measure the analytes in a mixture with the use of smaller amounts of material.

Based on chromatographic bed shape, chromatography is generally classified into column chromatography and planar chromatography. Basically, chromatography is classified into gas, liquid, and supercritical fluid chromatography (Figure 6.1). Liquid chromatography can be classified further based on properties such as size, charge, and affinity. To identify and characterize the resulting product, several identification methods such as Fourier transform infrared spectroscopy (FT-IR) for identifying functional groups, mass spectrometry (MS) for determining molecular weight, and nuclear magnetic resonance (NMR) spectroscopy for predicting the molecular structure, and so on, are employed. Data from all the methods are put together to predict the formed product.

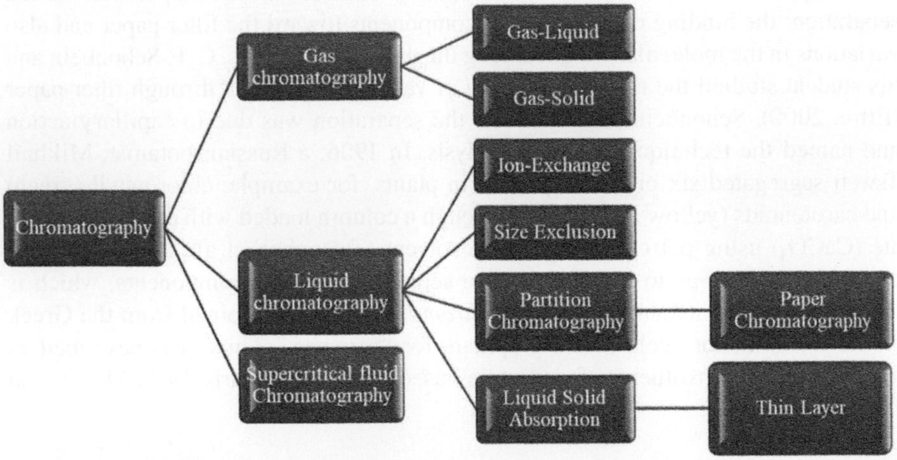

FIGURE 6.1 Branches of chromatography. (From Ettre, L.S., Evolution of liquid chromatography: A historical overview, in Cs. Horvath, Ed., *High-Performance Liquid Chromatography: Advances and Perspectives*, Vol. 1, Academic Press, New York, 1–74, 1980.)

6.1.3 Chromatographic Terms

The nomenclature of chromatography has been developed over the years, especially during the late 1950s to denote the characteristics of a chromatogram (Ettre and Sakodynskii, 1993). The *analyte* is the component(s) to be separated from the mixture. A bonded phase that is covalently linked to the support particles present in the inside wall of the column is termed as *stationary phase*. *Analytical chromatography* detects the presence and concentration of an analyte(s) in a sample. A *chromatogram* is the output displayed in the monitor of the chromatograph. Each peak or pattern on the chromatogram corresponds to a component(s) present in the mixture. The analyte is carried by the mobile phase solvent, called *eluent*, and *eluate* is the mobile phase that leaves the column. An *eluotropic series* is the order of solvents based on their eluting power. The *mobile phase* is the phase that flows in a defined direction. It may be a liquid (LC), a gas (GC), or a supercritical fluid (supercritical-fluid chromatography [SFC]). The sufficient quantities of a substance are purified using *preparative chromatography* for further use. The time taken by a particular analyte to pass through the entire system is termed *retention time*. The *sample* is the mixture to be separated in chromatography. The *solute* is the single component or mixture of components present in the sample. The *solvent* has the capacity to solubilize another substance completely. The *detector* refers to the instrument used to analyze the analytes qualitatively and quantitatively after separation. A typical chromatogram is shown in Figure 6.2.

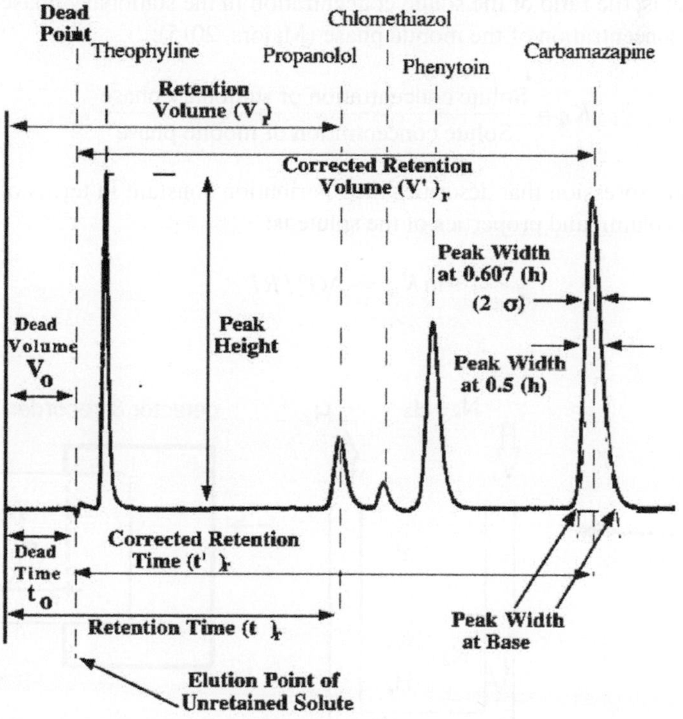

FIGURE 6.2 A typical chromatogram with its nomenclature. (From Scott, R.P.W. and Weissberger, A., *Contemporary Liquid Chromatography*, Wiley, New York: 1976.)

6.2 PRINCIPLE AND CONTROL OF SEPARATION

6.2.1 Principle of Chromatography

Chromatography is named commonly for the techniques which are based on the distribution of the target molecules between a mobile and a stationary phase (Figure 6.3). In a chromatography system, a liquid is pumped through a bed of particles packed in a column. The mobile phase comprises liquid, and the stationary phase comprises the packed particles. The mixture to be separated is injected into the mobile phase. The interaction between the stationary phase and target molecule is based on certain properties such as chemical nature and binding affinity, and thus separation is carried out as a function of time. The weaker the interaction, the faster is the transport of the analyte through the system, and vice versa. Slower transported molecules have stronger interactions. Thus, the retention time in the system is the parameter used to identify many analytes in a sample for a given set of conditions (Karger et al., 1973).

6.2.1.1 Distribution Coefficient

In practice, distribution takes place with the stationary and mobile phase inside the tube column packed with particulate matter. The *distribution constants* (K_D) are defined as the distribution of solute molecules between the mobile and stationary phases, that is, the ratio of the solute concentration in the stationary phase to that of the solute concentration of the mobile phase (Majors, 2015):

$$K_D = \frac{\text{Solute concentration of stationary phase}}{\text{Solute concentration of mobile phase}} \tag{6.1}$$

The typical expression that describes the distribution constant in terms of temperature of the column and properties of the solute is:

$$\ln K_D = -\Delta G^\circ / RT \tag{6.2}$$

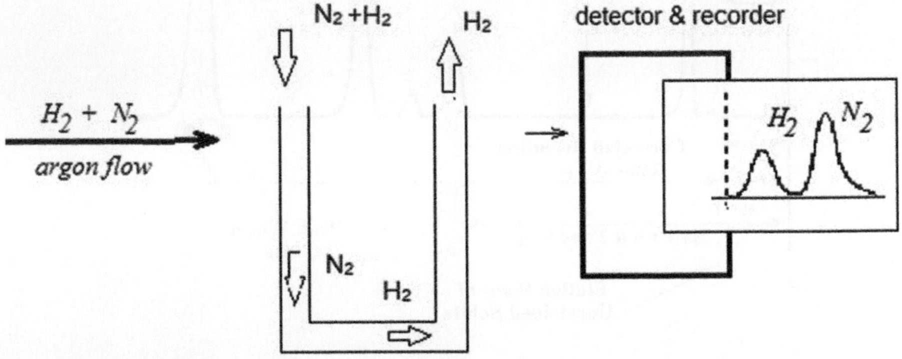

FIGURE 6.3 The principle of chromatographic analysis.

where, ΔG° is the change in the value of Gibbs free energy due to the evaporation of a molecule from the stationary phase, T is the temperature of the column, and R is the ideal gas constant.

6.2.1.2 Adsorption

Adsorption chromatography was the first chromatography system developed. The system comprises a stationary phase (solid) and a mobile phase (liquid or gas). During the twentieth century, pigments present in plants were adsorbed using a solid stationary phase of calcium carbonate and a liquid mobile phase of hydrocarbon. The different solutes carried by the solvent traveled different distances through the solid. Each solute is unique in achieving equilibrium between adsorption onto the solid surface and solubility in the liquid solvent (Heftmann, 1983). The components are separated into different bands based on the adsorbing power; that is, the components that are best adsorbed and are less soluble move slowly. Liquid chromatography containing a column packed with alumina or silica-gel is an example of adsorption chromatography (Figure 6.4).

6.2.1.3 Partition

In partition chromatography, a thin layer (or film) of nonvolatile liquid on the surface of an inert solid acts as the stationary phase. The mixture is distinguished using gas or liquid as the mobile phase (Martin, 1950). The solutes in the sample distribute between the mobile and the stationary phases. The component having stronger solubility in the mobile phase reaches the end of the chromatography column faster than the weaker soluble one. Paper chromatography is the best example of partition chromatography (Figure 6.5).

6.2.1.4 Ion Exchange

Ion exchange chromatography is more or less similar to partition chromatography. A resin-coated solid acts as the stationary phase (Gerberding and Byers, 1998). The resin is covalently bonded to different ions such as cations or anions, and it is paired electrostatically to ions of the opposite charge. The electrostatically and loosely bound ions are eluted along with the mobile phase first as other ions are strongly bonded. Softeners used in domestic water work on this principle (Figure 6.6).

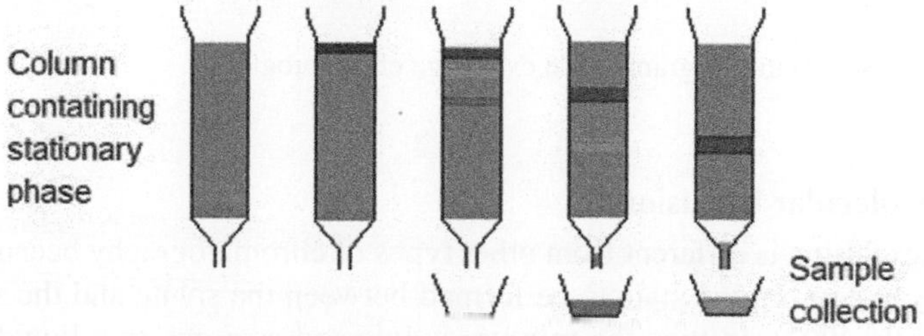

FIGURE 6.4 Schematic diagram of adsorption chromatography.

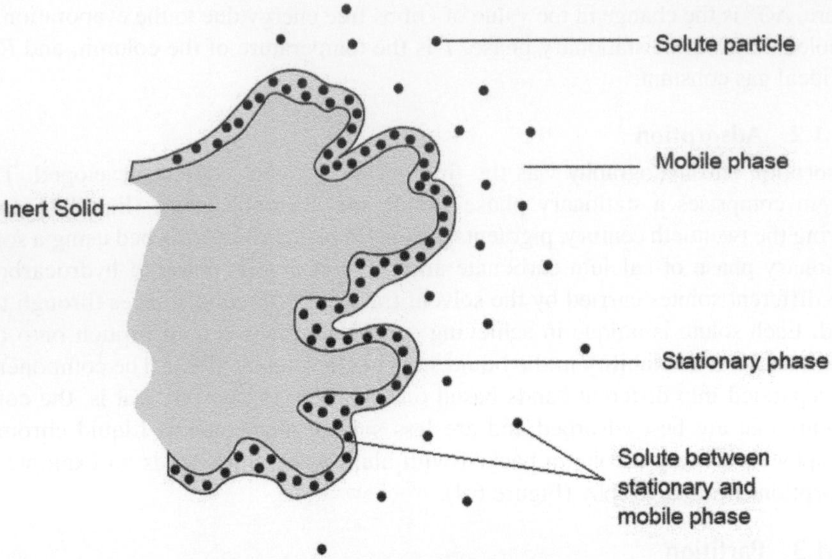

FIGURE 6.5 Schematic diagram of partition chromatography.

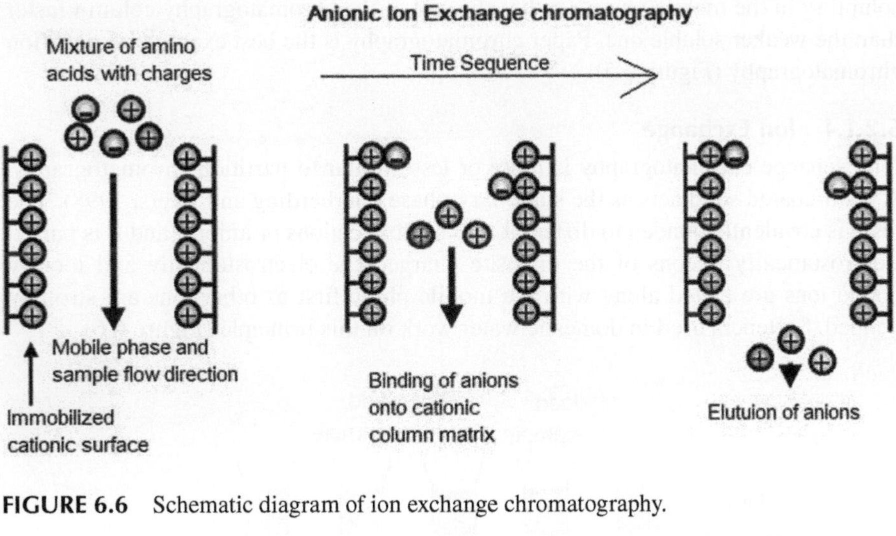

FIGURE 6.6 Schematic diagram of ion exchange chromatography.

6.2.1.5 Molecular Exclusion

Molecular exclusion is different from other types of chromatography because it does not require an equilibrium state to be formed between the solute and the stationary phase (Porath, 1997). In this, the mixture is injected as a gas or a liquid to travel

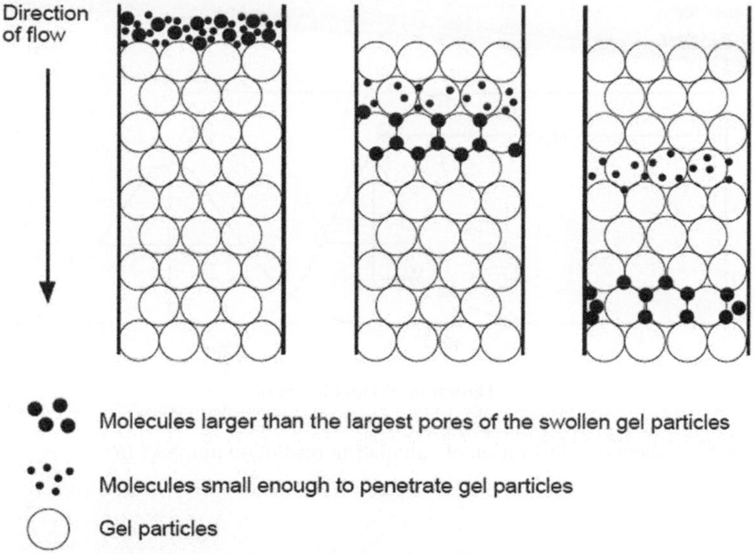

FIGURE 6.7 Schematic diagram of molecular exclusion chromatography.

through a porous gel. The pore size is designed so that it allows the solute particles to pass based on their size. The large particles pass through freely without any disturbances. The small particles are slowed down because they permeate through the gel and take longer get through the column. Thus, the particle size plays a vital role in the design of the column for the separation process (Figure 6.7).

6.2.2 Retention and Resolution in Chromatography

The *retention time* (t_R) is the time taken by the target molecule (analyte) to travel through the system. Dead time (t) is the time needed by the nonretained solute to travel through a stationary phase (Gilroy et al., 2003). The partition ratio or capacity factor (k') is the time required by a molecule to remain in the stationary phase relative to that of the mobile phase (Figure 6.8):

$$k' = \left(t_R - t_o\right) / t_o \tag{6.3}$$

The resolution number (R) is the term that illustrates the separation of two peaks (Caballero et al., 2015). Resolution numbers can be estimated using either Equation (6.4) or Equation (6.5):

$$R_s = \frac{1.18\left(t_{R2} - t_{R1}\right)}{\left(w_{h1} + w_{h2}\right)} \tag{6.4}$$

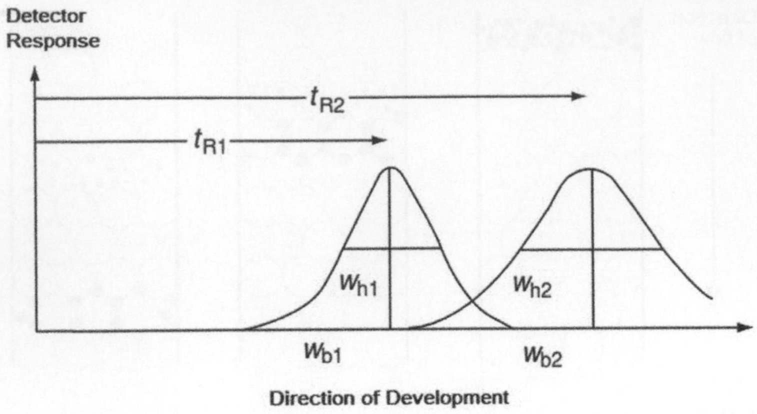

FIGURE 6.8 Schematic illustration of calculating resolution numbers (R).

$$R_s = 2(t_{R2} - t_{R1}) \big/ (w_{b1} + w_{b2}) \tag{6.5}$$

where t_{R1} and t_{R2} are the retention times of peaks 1 and 2, respectively; w_{h1} and w_{h2} are the peak widths at half height of peaks 1 and 2, respectively; and w_{b1} and w_{b2} are the peak widths at the base of peaks 1 and 2, respectively.

The theoretical plate number (N) plays a vital role in determining the separation capacity of a chromatographic column (Bose et al., 2013). The theoretical plate number can be determined by the following equation:

$$N = 5.545\left(t_R \big/ w_{0.5}\right)^2 \tag{6.6}$$

The theoretical plate number (N) is related to the height of the column (H) and can also be calculated using Equation (6.7):

$$N = \frac{L}{H} \tag{6.7}$$

where L is the length of the column packing (constant for a column).

The theoretical plate number indicates that resolving power can be improved by having more plates in the column. As the column has a definite length, the plate number can be increased by using thinner plates. It indicates that N and H are inversely proportional (Das and Mukherjee, 2011). For a given column, as N increases, H decreases. Thus, the following range of magnitudes exists for N and H:

Plate numbers: 100–1,00,000
Plate height: 0.1–0.001 mm or smaller

6.2.3 Components of a Chromatography System

The components of a chromatography system (Thammana, 2016) are depicted in Figure 6.9 and described below:

1. *Reservoir*: Bottle that holds mobile phase (buffer).
2. *Pump*: The buffer from the reservoir is pumped using one or two pumps. Various types of pumps are employed in chromatography system, depending generally on the pressure required to carry out the separation process.
3. *Mixer*: The buffer pumped from the reservoir is mixed from both pumps based on the requirement of the linear or step gradient.
4. *Column*: It is usually cylindrical made up of glass or steel.
5. *Detector*: The presence of analyte is detected by passing the eluate that comes out from the column and goes to the detector system. The process is based on different properties. Examples of various detectors are UV-visible detector (UVD), refractive index detector (RID), flame ionization detector (FID), and so on.
6. *Fractional collector*: A fractional collector is employed to collect the eluent corresponding to the peak as different fractions.
7. *Recorder*: It plots the characteristics of eluent in relation to its property measured in a detector.

6.2.4 Control of Separation

The best performance from HPLC techniques was obtained by optimizing the separation efficiency and dispersion of solutes. Ultimately, it is desirable to minimize individual peak widths, maximize individual peak heights, maximize the separation between analyte peaks, and minimize the analysis runtime. The Van Deemter equation hints at column parameters that can be controlled in order to achieve the desired chromatographic separation.

6.2.4.1 Van Deemter Equation

The characteristic rate equation that illustrates the chromatographic separation efficiency depending various physical column parameters is known as the Van Deemter equation. It incorporates the kinetic theory of chromatography to relate the

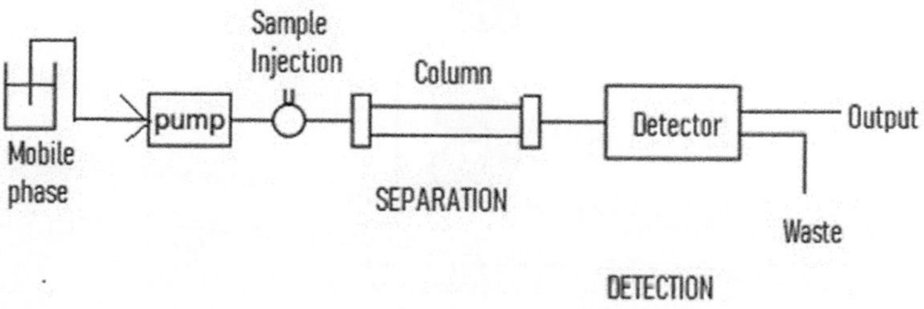

FIGURE 6.9 Components of a typical chromatographic system.

migration and dispersion of solute bands through a column packed with spherical particles in a constant mobile phase flow. The goal in understanding the separation process is to enhance column efficiency by understanding how to reduce band broadening. Several parameters have been highlighted in the Van Deemter equation for a chromatographic run that can be manipulated to reduce band broadening. According to the Van Deemter equation, band broadening may happen for the following reasons:

1. An analyte can take multiple paths to travel through the packed materials in a column
2. Molecular diffusion
3. Mass transfer effect between phases

Equation 6.1 describes the reasons of band broadening (major column parameters) and denote as height equivalent to theoretical plate (HETP) or H (Ohzeki et al., 1976):

$$H = 2\lambda d_p + \frac{2\gamma D_m}{u} + \frac{\left(1+6k'+11k'^2\right)d_p^{\,2}}{24\left(1+k'\right)^2}u + \frac{8}{\pi^2}\frac{k'd_f^{\,2}}{\left(1+k'\right)^2}u \tag{6.8}$$

The factors λ, γ, and ω do not vary significantly; k' is the capacity factor, d_f is the film thickness of the stationary phase, D_m is the coefficient of diffusion that illustrates the diffusion speed of the analyte in the mobile phase, d_p is the particle diameter in stationary phase, and u is the linear velocity. The fluidic explanations of the terms A, B and C are shown in Figure 6.10.

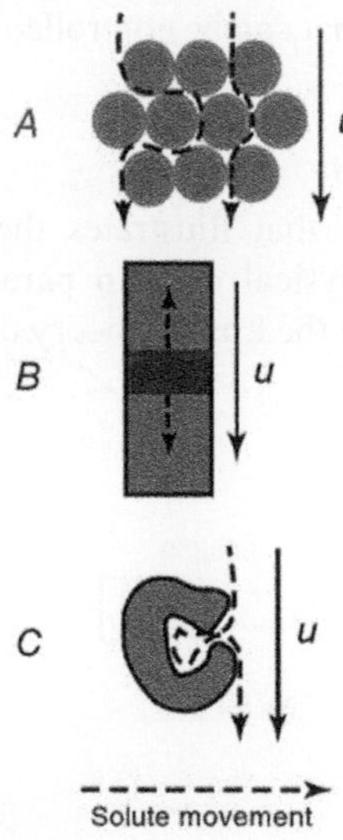

FIGURE 6.10 Physicochemical background of Van Deemter terms *A*, *B*, and *C*.

Equation (6.8) simplifies to the following:

$$H = A + \frac{B}{u} + Cu \tag{6.9}$$

HETP is a theoretical construct used to determine the effectiveness of a separation column. Greater efficiency of separation in a column is achieved by increasing the theoretical plate number, and it results in lowering the value of HETP. The *A* term describes eddy diffusion and it illustrates the different paths followed by a solute. The *B* term depicts the effect due to the molecular diffusion of the solute while passing through the column. The *C* term indicates the rate of mass transfer of the solute between the stationary phase and mobile phase (see Figure 6.10). The equation 6.9 assumes that the composition of the mobile phase is constant (isocratic elution). The theoretical graph in Figure 6.11 highlights the contribution of each term of the main equation (Hawkes, 1983). The resulting summed curve is the theoretical Van Deemter plot: a hyperbolic function that has a minimum value of plate height (or maximum efficiency) at a specific value of flow velocity (μ).

6.2.5 Key Parameters Influencing Chromatographic Separations

According to the Van Deemter equation (Equation 6.1), several parameters influence H. Recently, small particle size has been discussed as a tool for improving column efficiency. Although the size of the particles for column packing is a significant parameter, this is not the only factor that can be optimized to provide higher efficiency (Regueiro et al., 2014). For optimal chromatographic performance, the following parameters should be considered:

1. Particle size
2. Column dimensions
3. Column packing chemistries

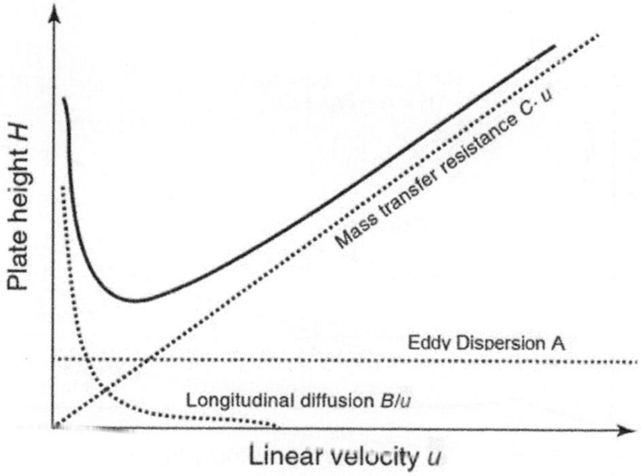

FIGURE 6.11 Theoretical Van Deemter plot.

4. Flow rate of mobile phase (linear velocity)
5. Mass transfer (stationary phase and mobile phase)
6. Gradient elution
7. Hardware improvements

6.2.5.1 Particle Size

Figure 6.12 shows that the *A* term is decreased effectively by reducing the diameter of particle size, thus lowering the theoretical HETP curve. The same HETP value can be achieved at a higher mobile phase linear velocity, resulting in faster separation. As a result, the speed of separation can be increased with smaller diameter particles, without loss of resolution (Vallverdú-Queralt et al., 2010).

6.2.5.2 Mass Transfer and Longitudinal Diffusion

The increase in the speed of separation can be observed by manipulating the *B* and *C* terms of the Van Deemter equation (Figure 6.13). Techniques like reducing the size of the particles, turbulent flow chromatography, capillary columns, and monolithic phases all take advantage of changes in the mass transfer and flow dynamics within the column to flatten the Van Deemter curve, thus allowing increased efficiency at higher velocity (Stauffer et al., 2008).

6.2.5.3 Gradient Elution

Gradient elution changes everything. The basic Van Deemter equation can be applied for separations involving isocratic elution. Gradient elution added another dimension to the Van Deemter equation, basically allowing the B diffusion term and the partition value *k*′ (Equation 6.1) to vary with time (mobile phase composition). This brings in a whole new set of terms to express band broadening associated with the composition of the mobile phase and the distribution of solutes in the column. In this case, the contribution to band broadening (and the reduction in column efficiency) from imprecision in the gradient becomes more significant than the *B* and *C* terms. As a result, the ability to generate a high precision gradient can significantly

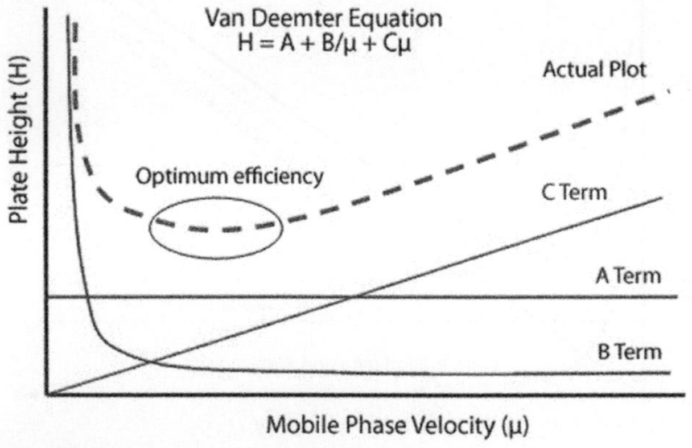

FIGURE 6.12 Effect of particle size on Van Deemter plot.

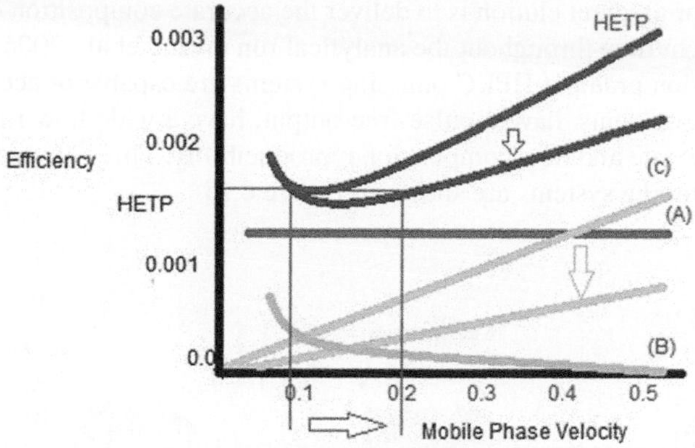

FIGURE 6.13 Effect of changes to the mass transfer chromatography on Van Deemter plot.

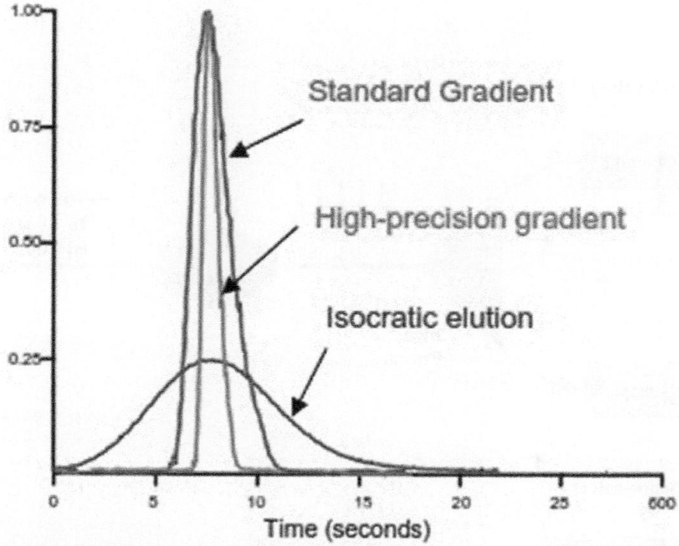

FIGURE 6.14 Effect of high precision gradient on peak solution.

affect column efficiency. Figure 6.14 shows how a high-precision gradient elution can influence peak width on a typical column (Dolan and Snyder, 2012).

6.2.5.4 Hardware Improvements

To improve the separation process with the precise delivery of solvents, several pumping systems have been introduced over the years. Administering one solvent of constant composition is meant as an isocratic elution, and it is typically used only for the applications that need no separation, or for method development. More typically, separation efficiency necessitates improvement by gradient elution. The main aim of a pump used for gradient elution is to deliver the accurate composition of the mobile phase at the right time throughout the analytical run (Wang et al., 2006).

High-precision gradient HPLC pumping systems are capable of accurate mixing of two or more solvents, have a pulse-free output, have a wide flow rate range, and control the flow rate and flow composition reproducibility. Three characteristic types of gradient pumping systems are shown in Figure 6.15.

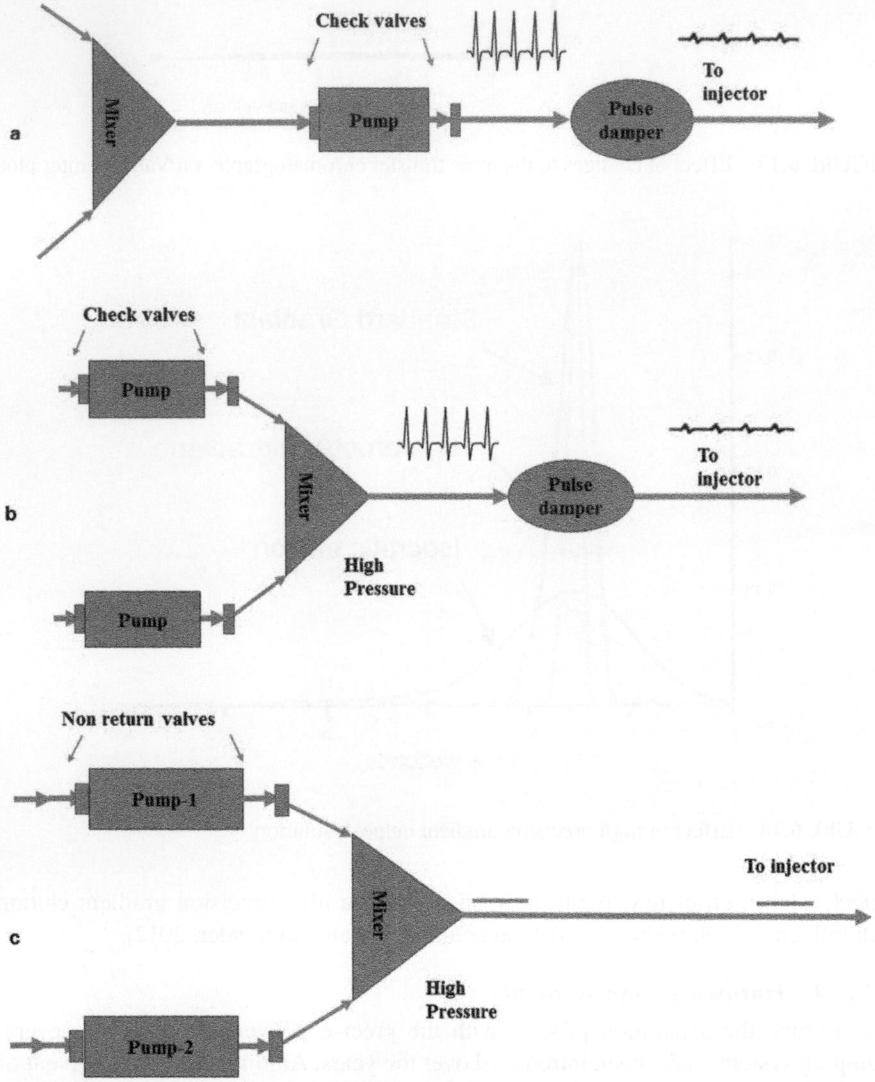

FIGURE 6.15 Typical binary gradient pump configurations: (a) reciprocating pump (low-pressure mixing), (b) reciprocating pump (high-pressure mixing), and (c) dual syringe high-pressure mixing.

TABLE 6.1
Critical Factors to Be Considered for Improving Chromatographic Operation

Factor	Advantages	Disadvantages
Longer column length	Better separation, more theoretical plates	Longer runtimes, higher back pressure
Smaller particle size	Higher efficiency, shorter runtimes	Higher back pressure, possible column robustness problems
Monolithic stationary phases	Higher efficiency, shorter runtimes	Higher flow rates
Smaller diameter columns	Higher efficiency, better separations in shorter time	Smaller column loading, possible robustness problems
Higher flow rates	Faster runtimes	Reduced efficiency, higher back pressures, difficult interface for ESI
Better gradient precision	More efficient, faster separations	Requires improved pumping technology
Higher pump pressure	Faster runs at high efficiency	Requires improved pumps and valves, possible column stability issues
Higher temperature	Reduced back pressure	Column, analyte stability and reproducibility issues

6.2.5.5 Parameter Summary

More than one factor must be considered in order to improve chromatographic performance. A series of steps can be taken, including the use of small particle sizes and/or different column dimensions and packing chemistries and/or high precision/resolution gradient elutions. Table 6.1 summarizes some critical factors to consider in order to achieve significant improvements in chromatographic resolution, sample throughput, and sensitivity.

6.3 HIGH-PERFORMANCE LIQUID CHROMATOGRAPHY

6.3.1 Introduction

In 1941, Martin and Synge stated, "Smallest HETP across the length of the column can be achieved by using very small particles and this in turn results in a high pressure difference." The current practice of LC is in relation to this statement (Rogatsky, 2016).

High-performance liquid chromatography (HPLC) is the most common preparative or analytical tool used in many areas of research. HPLC is employed for:

1. Qualitative and quantitative analysis of unknown mixtures.
2. Separation of components from the mixtures for future analysis—preparative HPLC.

The particles in the stationary phase are very small in HPLC. A typical particle size for HPLC is 5 μm compared to 60 μm for column chromatography. The flow rate of solvent of several mL/min through such dense packed material in a stainless steel tube requires high pressure, up to several thousand pounds per square inch (psi). The tiny stationary phase particles are chosen because the smaller the particle size, the higher the surface area, which improves the interaction between the analyte and the stationary phase, which in turn results in better separation (Majors, 2019).

The components of HPLC include a reservoir, pump, injector, mixer, the column, and a detector or recorder. These components are interconnected in series via a steel tube (Figure 6.16). The pump is meant for controlling the flow rate of solvent through the system. With the controlled flow rate, the solvent reaches the injector. Then, along with the injected sample, it travels through the column and finally reaches the detector or recorder. The injector limits the volume of sample to be injected and thus creates a convenient flow of sample onto the column. The detector identifies the particular analyte and displays it as a peak. Finally, the sample, along with the solvent, elutes or leaves the column. The separated components are fractionated individually after collection in a fraction collector for further analysis (Zotou, 2012). The retention time and the peak area in the chromatogram are the two significant parameters to understand about each analyte. Figure 6.17 shows the chromatogram of diet soft drinks. Figure 6.18 shows the large surface area particles of silica (the stationary phase in HPLC) that have a "greasy" coating on the surface.

6.3.1.1 Pros and Cons of HPLC

An appropriate selection of column, solvent, and detector is necessary for more accurate separation of target molecules from others. The whole HPLC system can also be used for simultaneous identification and quantification of individual components present in a mixture, such as amino acids, drug metabolites, and so on. This is the greatest advantage of HPLC over radioimmunoassay because radioimmunoassay

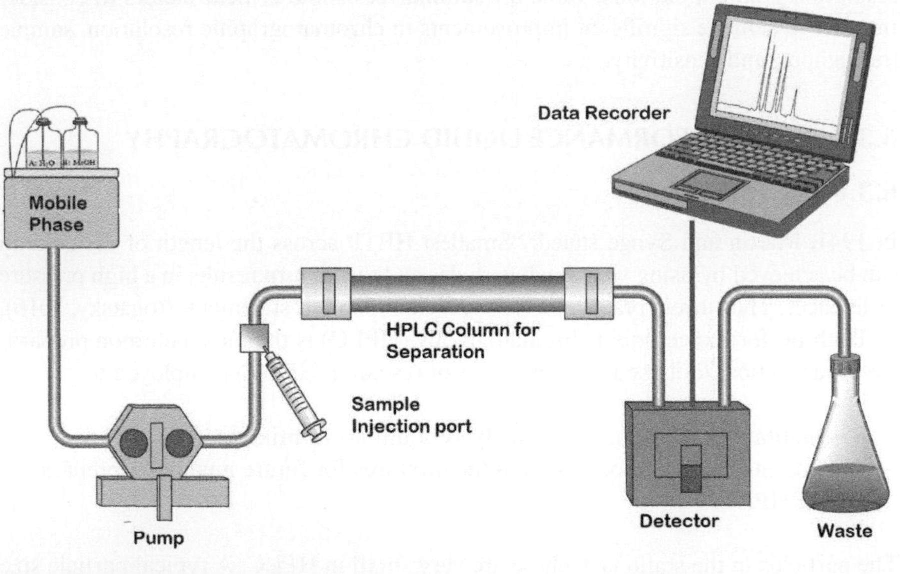

FIGURE 6.16 A typical HPLC setup.

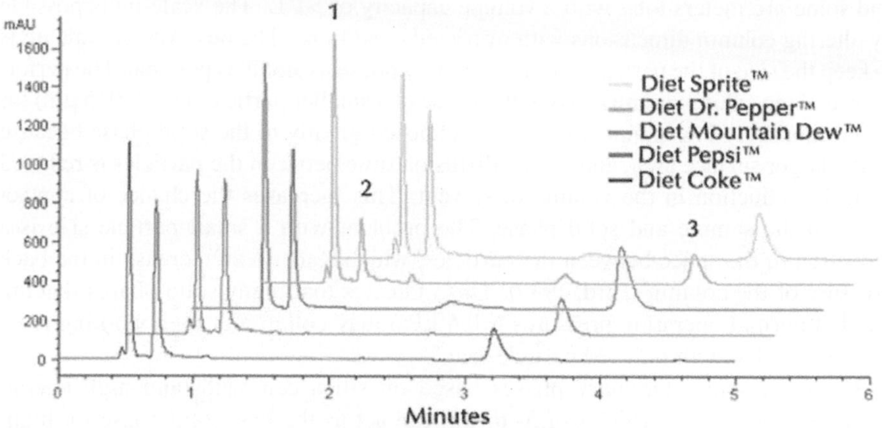

FIGURE 6.17 Chromatogram of diet soft drinks.

FIGURE 6.18 A typical silica particle as stationary phase for HPLC.

requires separate assays. Further, HPLC enables the identification of many unknown compound(s) in a sample, whereas only target compound(s) can be detected in radioimmunoassay. Unlike radioimmunoassay, HPLC can be applied readily for analyzing a large amount of a sample. With the help of automated systems for sample injection and analyte collection, a good throughput of HPLC can be achieved. Many companies are now manufacturing such systems. The investment for developing such systems is quite high, but the running costs are low. The sensitiveness of HPLC is a drawback because a small interference by sample or solvent contaminants may result in different outputs and makes the HPLC method unusable in routine analysis. Preparing the samples and solvents with great care and maintaining purity may solve the problem.

6.3.2 Principle

The characteristic feature of any chromatography is the ability of the column to retain a sample molecule effectively. The sample should have adequate time in contacting the stationary phase. This can be achieved by making the column bigger Some columns are designed few centimeters long with a volume capacity of <10 mL, and some are meters long with a volume capacity of >1 L. The scale-up is possible by altering column dimensions with restricted conditions. The best way to scale up is to keep the size of the particles of the stationary phase as small as possible. The performance of the column is enhanced with the use of a smaller particle size (250–5 μm) for two reasons: (1) The sample molecule is exposed greatly to the solid phase because of the larger surface area, and (2) the diffusion time between the particles is reduced with the reduction in the volume of solvent. This increases the chance of contact between the sample and solid phase. The problem with a small particle size is a reduction in the space between the particles, which leads to an increase in the back pressure of the column (Bird, 1989). This indicates that many solid phases that are good at normal operating pressures (<0.5 kPa) may collapse at high working pressures (3–5 kPa) with reduced particle size.

However, some stationary phases based on silica can withstand high operating pressures. Silica particles of 5–10 μm can act as the best solid phase for high-performance chromatography. Silica is highly versatile and acceptable for chemical modification on its surface, which makes it best for ion exchange and a great reverse phase medium. Physical modifications of silica by creating tiny pores on its surface make it a good choice for retarding smaller molecules in the case of size exclusion chromatography. The flexibility of silica for modification makes it a good candidate for use in the separation of most molecules. Thus, columns packed with silica-based stationary phases can be applied for most HPLC methods (Thammana, 2016).

Column efficiency is a factor that indicates the performance of the columns. It can be calculated by the following equation:

$$n = 16\,(t_r^2)/(t_w) \qquad (6.10)$$

where n indicates efficiency of the column, t_r is the retention time of a sample molecule, and t_w is the width of the peak. The n values would be <500 for ordinary columns, but they can easily exceed 10,000 for high-performance columns. Figure 6.19 shows how chromatographic separation works.

Retention time (qualitative) and the peak area (quantitative) are the primary pieces of information needed for each sample. The retention time for the standard and the sample should be same (Warren and Vella, 1995).

6.3.3 Application

HPLC has a significant range of applications in the field of clinical research. The variety of molecules that are commonly detected by this method are discussed here. Longer-term plasma glucose control can be monitored in diabetic patients by estimating glycated hemoglobin, type 1c. A method based on charge can be employed for the separation because at neutral pH, all the 1a, 1b, and 1c forms of glycated hemoglobin are less positively charged than is normal hemoglobin. Chromatography based on cation exchange columns and also electrophoresis methods can be applied to separate glycated from nonglycated hemoglobin. Major interferences may occur in the separation process because of other forms of hemoglobin, such as HbF, HbS, and HbC, associated with hemoglobinopathies present

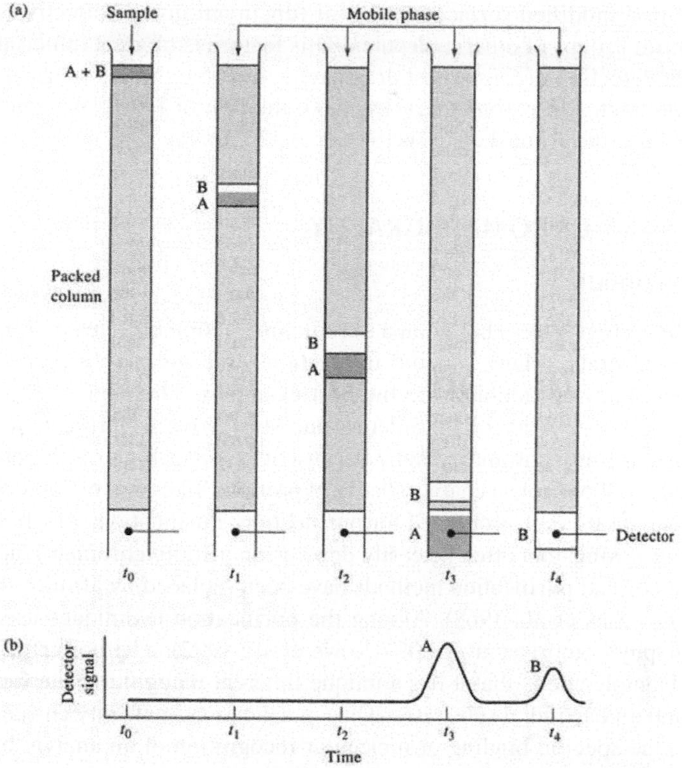

FIGURE 6.19 (a) Components A and B are separated from a mixture by column elution chromatography. (b) The output shown in the detector is based on the elution of samples A and B in relation to part (a) of the figure. (From Skoog, D. A. et al., *Principles of Instrumental Analysis*, Published in 1998 in Belmont (Calif.) by Brooks/Cole, 5th ed., Saunders College Publishing, Philadelphia, PA, 1998.)

along with the glycated hemoglobins. Cation exchange HPLC provides solution to this problem because it can resolve all the subtypes of glycated hemoglobin. This method shows excellent precision with rapid separation.

Congenital metabolic disorders can be investigated using HPLC because of its high resolution power. Conventional chromatography such as paper chromatography and thin-layer chromatography can be applied for separating amino acids in plasma or urine. But because of its poor resolution and complications in quantification, ion exchange HPLC methods can be employed. Increasing the solubility of 27 amino acids in organic media enables its separation using reverse phase HPLC. Leucine, isoleucine, and valine, which are indicators of maple syrup urine disease, can be detected using thin-layer or paper chromatography. A high-resolution technique like HPLC can be employed to separate phenylalanine because of its vigorous concentration difference (Petrova and Sauer, 2017).

The application of radioimmunoassay fails in the case of monitoring any drug that does not have an antibody available for the drug. In this case, HPLC can be applied more effectively and also for the simultaneous quantification of a mixture of drugs.

Silica with a modified surface and chiral functional groups effectively binds to selective chiral groups of other molecules. This feature is of great importance in the area of drug detection because most drugs are active in either of their chiral forms. Rapid developments in chromatography make the role of HPLC inevitable in both research and routine clinical analysis (Sun et al., 2009).

6.4 AFFINITY CHROMATOGRAPHY

6.4.1 Introduction

Affinity chromatography is based on a specific interaction between an immobilized ligand and substrate. Many specific interactions like antibody-antigen, enzyme-substrate, and enzyme-inhibitor are the best examples. The degree of purification depends on the specificity of the interaction. Most protein purification processes mainly utilize affinity chromatography because it provides high resolution, selectivity, and capacity. The typical characteristics of proteins, like their biological structure and function, are used as properties for purification. Purification, which is complex and time consuming, can often be easily done using affinity chromatography.

Many traditional purification methods have been replaced by affinity chromatography (Cuatrecasas et al., 1968). Among the purification techniques used, affinity chromatography comprises over 60% (Lowe et al., 1992). The powerful feature of any given biomolecule is that it has a unique inherent recognition site for the binding of natural and artificial substrates. Thus, purification in affinity chromatography is based on the specific binding or molecular recognition of an analyte by a ligand bound to a solid matrix. Affinity purification follows three major steps:

1. The crude sample is incubated with the affinity support specific binding of the target molecule to the immobilized ligand.
2. The unbound components are washed away from the support.
3. By altering the buffer condition, the target molecule is eluted from the immobilized ligand (Roque and Lowe, 2008).

Emil Starkenstein published an article in 1910 about resolving macromolecule complexes based on their specific interactions with an immobilized substrate. This article describes the effect of chloride on the enzymatic activity of liver α-amylase, and it encouraged several researchers to work on it (Arsenis and Mccormick, 1964; Bautz and Hall, 1962; Campbell et al., 1951). In 1968, the name, affinity chromatography, was introduced by Pedro Cuatecasas, Chris Anfinsen, and Meir Wilchek in an article that dealt with the purification of enzymes using immobilized substrates and inhibitors (Wang et al., 2017). Other articles illustrated the use of sepharose matrix by activating it using a cyanogen bromide (CNBr) reaction (Axén et al., 1967) and the applicability of spacer arm to lessen steric hindrance (Clonis et al., 2000).

Affinity chromatography is still evolving and plays a vital role in genomics, proteomics, and metabolomics. The powerful feature of affinity chromatography allows researchers to study protein-protein interaction and degradation, and post-translational modification. The attachment of reversed phase affinity chromatography with mass spectrometry has enabled researchers to discover many protein biomarkers.

6.4.2 Principle

Affinity chromatography is based mainly on the reversible affinity of the target protein to the ligand coated on to column matrix. A lock-and-key model best describes the affinity binding of the protein and ligand. Each protein has a unique structure key called a recognition site that can be used for binding to select locks called ligands. The specific interaction between the target protein and the selected ligand must be certain characteristics, such as specific and reversible. Typical steps involved in affinity purification are shown in Figure 6.20. Initially, the conditions are set up so that the samples have maximum binding affinity toward the ligand. The nonbounded samples are washed away by using washing buffer, leaving the target protein bound to the column matrix. Next, a desorption step elutes the target molecules that are bound to the column matrix by using either of the following two steps: (1) using a specific competitive ligand or (2) changing the condition of the column (for example, ionic strength, pH, or polarity) (Zachariou, 2008). After the elution step, the purified target protein can be obtained in a concentrated form.

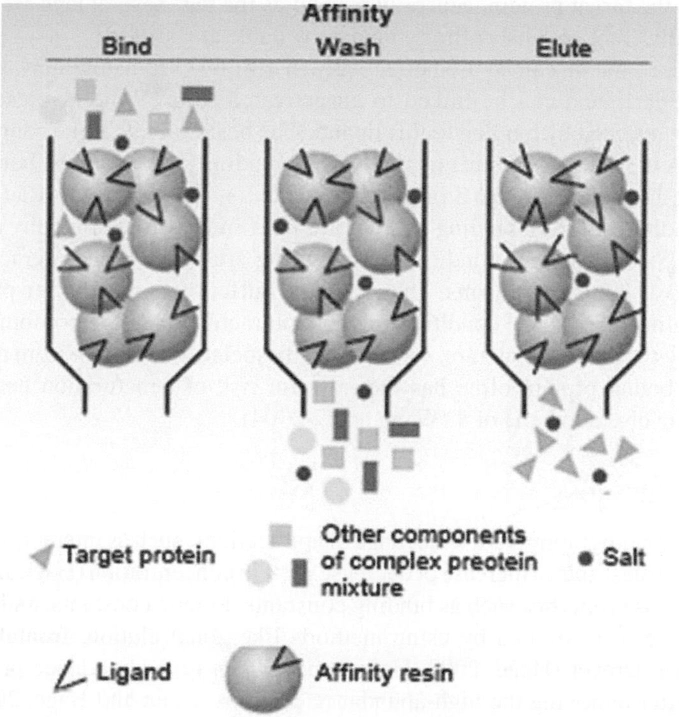

FIGURE 6.20 Schematic diagram of affinity chromatography.

6.4.2.1 Affinity-Tagged Purification

Affinity-tagged protein purification is carried out in three step process. Protein samples that contain the affinity-tag sequence, generally Polyhistidine (His) and Glutathione S-transferase (GST), selectively bind to the specific ligand in the first step. In the second step, a washing buffer is added to remove the contaminants and nonbound protein samples. In the third step, the elution buffer is added to elute the bound protein in pure form.

Affinity tags have many advantages. The specific feature of immobilized metal affinity chromatography (IMAC) is the interaction between the affinity tags and the immobilized metals. For example, affinity tags such as His and GST have good selectivity in binding with transition metals like Ni^{2+} immobilized to the ligand; the tagged protein bound to the column matrix can be eluted selectively with imidazole. Native conditions are used for purification of proteins that are tagged with an enzymatically active GST. Native or denaturing conditions can be employed for the purification of polyhistidine-tagged proteins. Salts and other foreign substances are removed by exchanging buffer during desalting to ease the process downstream. The buffer exchange can be done using various desalting techniques like dialysis, size exclusion chromatography, and ultrafiltration.

Affinity chromatography can be done using one of the following two methods. In the first method, the protein uses a set of amino acid sequence as its binding site. The bilirubin sequence in albumin protein is the specific binding site for Affi-Gel Blue support. In the second method, an engineered special amino acid sequence is attached to the target protein, generally termed as the tag. Several tags are available commercially. Table 6.2 lists affinity media available at Bio-Rad.

Customized media can be used effectively in affinity chromatography. For example, a specific ligand can be linked to an activated resin. Proteins present in the samples are retained by binding to this ligand. The best example is the coupling linkage of DNA to beads. Using this method, DNA-binding protein can be bounded and purified on the media. Table 6.3 lists activated media available at Bio-Rad.

The conditions during binding and elution in affinity chromatography can differ to a great extent. The best conditions for binding are the binding interactions that occur in most cellular organisms. Therefore, the buffer of choice is often phosphate-buffered saline (PBS). The conditions for the interactions that are not found in vivo may modify the protein structure, causing it to dissociate the protein from the ligand. Elution of bound protein often has the inherent risk of denaturation because few processes involve a low pH of 4 (Wang et al., 2004).

6.4.3 Applications

Affinity chromatography has a wide range of applications, such as interaction between protein and drugs, and an increase or decrease protein concentration (Hage and Austin, 2000). Various properties such as binding constants, kinetic constants, and allosteric interactions can be studied by using methods like zonal elution, frontal analysis, and Hummel-Dreyer (Hage, 2002; Hage et al., 2011). Low-abundance proteins are analyzed after removing the high-abundance proteins (Chen and Hage, 2004). This washing improves the concentration of low-abundance proteins and thereby eases their identification and quantification (Tu et al., 2010). Affinity chromatography has

TABLE 6.2
Ready-to-Use Affinity Media Selection Guide

	Matrix	Functional Group	Specificity	Capacity	Working pH	Pressure Limit	Applications
Nuvia™ IMAC	High capacity, pressure-stable polymer based on UNOsphere™ beads	NTA charged with Ni^{2+}	Histidine	≥40 mg/mL	2–14	45 psi (3.1 bar)	Purification of recombinant histidine-tagged proteins; can be charged with other transition metals
Profinity™ IMAC	Pressure-stable polymer based on UNOsphere beads	IDA, provided charged with Ni^{2+} and uncharged	Histidine	≥15 mg/mL	1–14	100 psi (6.8 bar)	Purification of recombinant proteins tagged with histidine; can be charged with other transition metals
Profinity™ GST	Pressure-stable polymer based on UNOsphere beads	Immobilized glutathione	Proteins tagged with GST	≥10 mg/mL	1–14	45 psi (3.1 bar)	Purification of recombinant proteins tagged with GST
Profinity eXact™	Crosslinked 6% agarose	Subtilisin protease	Subtilisin prodomain	≥3 mg/mL tag-free protein	2–10	10 psi (minus system pressure)	Generation of native, tag-free protein by on-column purification and cleavage
Affi-Gel® Protein A	Crosslinked agarose	Protein A 2 mg/mL	IgG		2–10	15 psi (1 bar)	Purification of IgG from ascites, serum, and culture fluid; with MAPS buffer system, purification of 10 mg mouse IgG per mL of gel is possible
Affi-Prep® Protein A	Pressure-stable polymer	Protein A 2 mg/mL	IgG		2–10	1,000 psi (70 bar)	Purified IgG from ascites, serum, and culture fluid; pressure-stable support for process-scale applications
Affi-Gel® Blue	Crosslinked agarose	Cibacron Blue F3GA 1.9 mg/mL	Albumin; general	≥11 mg/mL	2–10	15 psi (1 bar)	Binds many nucleotide-requiring enzymes, albumin, and other proteins

(Continued)

TABLE 6.2 (*Continued*)
Ready-to-Use Affinity Media Selection Guide

	Matrix	Functional Group	Specificity	Capacity	Working pH	Pressure Limit	Applications
DEAE Affi-Gel®Blue	Crosslinked agarose	Cibacron Blue F3GA and DEAE	Albumin and serum protiens	0.14 mL serum/mL gel	2–10	15 psi (1 bar)	Purifies protease-free IgG from ascites, serum, and culture fluid with minimal sample preparation
CM Affi-Gel®Blue	Crosslinked agarose	Cibacron Blue F3GA and CM	Albumin and serum protiens	0.17–0.5 mL serum/mL gel	2–11	15 psi (1 bar)	Produces albumin- and protease-free antibody preparation from serum without prior dialysis
Affi-Prep®Polymyxin	Pressure-stable polymer	Polymyxin 2–4 mg/mL	Endotoxins	>5 mg/mL	2–10	1,000 psi (70 bar)	Endotoxin removal
Affi-Gel®Boronate	Polyacrylamide gel	Boronate 1.05 ± 0.15 meq/g	cis-diols	130 µmol sorbitol/mL	2–10	15 psi (1 bar)	Adsorption of cis-hydroxyl–containing molecules, including sugars, nucleotides, and glycopeptides

Source: Bio-Rad, Affinity Chromatography | LSR | Bio-Rad, http://www.bio-rad.com/en-us/applications-technologies/affinity-chromatography?ID=LUSMJIDN, 2019.

TABLE 6.3
List of Activated Media for Ligand Immobilization

	Matrix	Functional Group	Specificity	Capacity	Working pH	Pressure Limit	Applications
Profinity™ Epoxide	Pressure-stable polymer based on UNOsphere beads	Epoxy group	Nucleophiles; amini, thiol, –COOH	36–40 mg/mL IgG	1–14	Up to 80 psi (5.5 bar)	Activated matrix for the immobilization of various ligands (examples, StrepTactin, protein A, and immunoglobulins)
Affi-Gel®10	Crosslinked agarose	N-hydroxy-succinimide ≥10 μmol/mL	$-NH_2$	35 mg/mL	3–11	15 psi (1 bar)	For coupling proteins with pI 6.5–11
Affi-Gel®15	Crosslinked agarose	N-hydroxy-succinimide ≥9 μmol/mL	$-NH_2$	35 mg/mL	3–11	15 psi (1 bar)	For coupling proteins with pI < 6.5
Affi-Gel®Hz	Crosslinked agarose	Hydrazide	Oxidized carbohydrates	1–5 mg/mL	2–10	15 psi (1 bar)	Immobilization of immunoglobulins and other glycoproteins via carbohydrate molecules
Affinity Media Using Carbodiimide Activation							
Affi-Gel®102	Crosslinked agarose	$-NH_2$ 16 ± 4 meq/mL	–COOH	40 mg	2–11	15 psi (1 bar)	Carbodiimide coupling of carboxyl-containing ligand

Source: Bio-Rad, Affinity Chromatography | LSR | Bio-Rad, http://www.bio-rad.com/en-us/applications-technologies/affinity-chromatography?ID=LUSMJIDN, 2019.

been developed as a powerful tool in the "omics" fields: for example, proteomics, metabolomics, and genomics, and attempts to develop high-throughput screening methods for the identification and quantification of potential drugs.

Immobilization of antibodies can be done through two methods: covalent and adsorption. Several amine groups such as tresyl chloride, *N,N'*-carbonyldiimidazole, and cyanogen bromide present in the solid support favor the covalent binding of antibodies. Free amine groups have the ability to react with groups like aldehyde or epoxy groups present on the activated support. The problem with random immobilization such as blocking of antibody binding sites, multisite attachment, or steric hindrance can be overcome by using site-specific covalent immobilization. It involves the conversion of carbohydrate groups in the Fc region of the antibody to produce aldehyde residues for reacting with amine groups (Ruhn et al., 1994). The adsorption of antibodies onto secondary ligands can also be done; for example, antibodies can bind with biotin through the reaction between the hyrazide group of biotin and the carbohydrate groups of the Fc region of the antibody (Hermanson, 1992). Thus, the produced biotinylated antibody can be adsorbed through affinity onto the surface of avidin or streptavidin. These types of immobilization are commercially available as bionylation kits. Antibodies can also adsorb directly onto a protein based on its specific affinity. For permanent immobilization, groups like carbodiimide or dimethyl pimelimidate present on the solid support can be utilized for adsorbing antibodies (Phillips et al., 1985).

The features of protein post-translational modification such as glycosylation can be done using the powerful tool of lectin affinity chromatography. Lectins are classified as carbohydrate-binding proteins and exhibit affinity to monosaccharides such as mannose, galactose, fucose, and *N*-acetylglucosamine (Sharon, 1998). In this affinity technique, glycosylated protein is bound to sugar groups (*N*-linked or *O*-linked) present in the immobilized lectin. Many types of lectin such as Concanavaline A (Con A), wheat germ agglutinin (WGA), and sepharose are commercially available for glycoprotein purification.

6.5 SIZE EXCLUSION CHROMATOGRAPHY

6.5.1 Introduction

Size exclusion chromatography (SEC) is employed for separating a mixture that has components of different particle sizes. Thus, the geometry of the molecules, such as shape and size, plays a major role in separation using this technique. Colin R Ruthven and Grant Henry Lathe were the first to apply this technique with starch gels as the matrix for separation of analytes of different sizes. Starch gels were replaced with dextran gels by Per Flodin and Jerker Porath. The gel filtration matrix includes polyacrylamide and agarose (Patil et al., 1970).

6.5.2 Principle

SEC is based on the principle of molecular size differences between the components present in a mixture. The components are separated by the differential exclusion as they travel through a stationary phase matrix containing polymeric gels or beads of heteroporous in nature. Each solute molecule has unique permeation rates into the interior of the gel particles, and this guides the separation process. SEC deals with the gentle interaction of the sample with the matrix, enabling high retention of biomolecular activity. SEC can be called by other names based on the material being separated: gel filtration chromatography (GFC) for separating biomolecules in aqueous systems, and gel permeation chromatography (GPC) for separating organic polymers in nonaqueous systems (Patil et al., 1970).

SEC consists of a column packed with heteroporous gel particles as the matrix and equilibrated with a suitable mobile phase (Figure 6.21). The particles of different size take different paths to travel through the column. The molecules larger than the pore size are excluded completely from entering the pores; they travel through the space present between the porous gel matrix or beads, elute, and are detected first. The molecules smaller than the pore size travel through the pores and are distributed in both the mobile and stationary phase; thus, these elute next at a slower rate (Determann, 1969).

SEC is based on the concept of critical molecular mass. The particles with values larger than or equal to the critical molecular mass are excluded completely

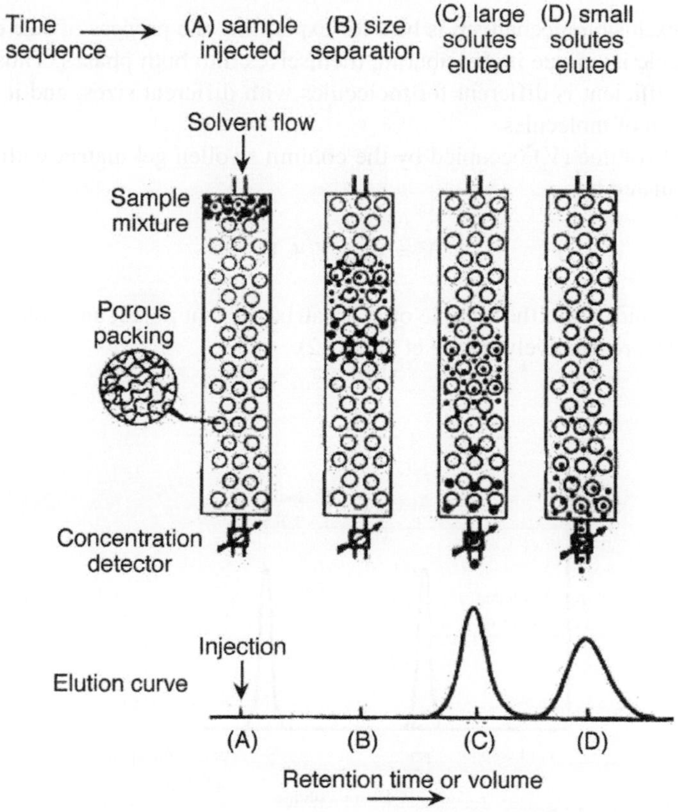

FIGURE 6.21 Schematic diagram of size exclusion chromatography.

from the polymeric gel matrix, whereas the molecules having molecular mass lower than the critical molecular mass travel completely through the pores of the column matrix. The molecules between these two extreme ranges of critical molecular mass are eluted through both the excluded and included volumes (Figure 6.22). Thus, the included and excluded volumes become the important parameters in designing a gel filtration bed matrix for protein purification. For example, the Sephadex G 75 has a volume range of 3–80, which means that the included and excluded volumes are 3 and 80 kDa, respectively. Generally, SEC is used for desalting purposes as all the proteins (>5 kDa) land in the excluded volume and the salts land in the included volume (Ó'Fágáin et al., 2011). The column length determines the capacity of resolution. A better resolution is obtained with longer columns. The excluded and included volumes correspond to one-third and two-thirds of the column volume, respectively.

Size exclusion is a type of liquid-liquid partition chromatography as the solute molecules are distributed between the two phases:

1. Liquid inside the pores of gel matrix
2. Liquid present in the void volume

The steric exclusion mechanism is best for explaining the process of size exclusion. Each molecule is unique in distributing themselves into both phases. Thus, the distribution coefficient is different for molecules with different sizes, and it results in the separation of molecules.

The total volume (V_t) occupied by the column swollen gel matrix with a solvent can be calculated by:

$$V_t = V_g + V_i + V_o \tag{6.11}$$

where V_g, V_i, and V_o are the volume of gel matrix, solvent pores, and volume outside the gel matrix, respectively (Hong et al., 2012).

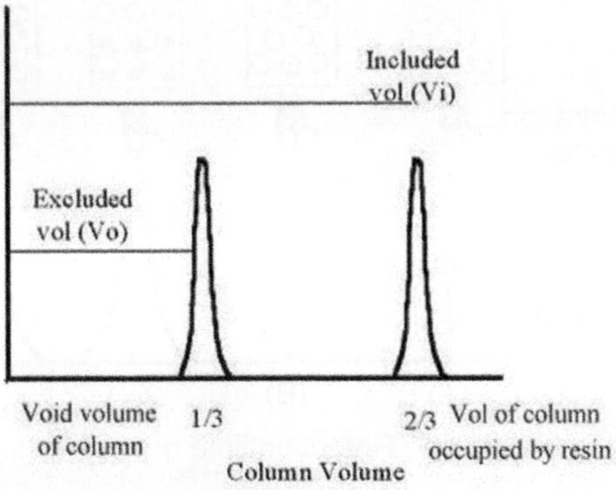

FIGURE 6.22 Schematic explanation of principle of size exclusion chromatography.

The retention volume (V_R) during mixing or diffusion can be calculated by:

$$V_R = V_{(int.)} + K_d\, V_{(int.)} \tag{6.12}$$

where distribution coefficient (K_d) is given by:

$$K_d = V_{i(acc)}/V_{(total)} \tag{6.13}$$

where $V_{i(acc)}$, $V_{(total)}$, and $V_{(int.)}$ represent the accessible pore volume, total pore volume, and the interstitial volume, respectively.

The secondary exclusion mechanism is also used to explain the mechanism of SEC. According to the mechanism, a sample containing both small and large molecules separate based on their diffusion into the porous gel matrix. The small molecules diffuse into the pores of the gel, whereas the large molecules travel outside the beads. Thus, the separation takes place between small and large molecules (Werner et al., 1990).

6.5.3 Application

SEC is used successfully for the separation of sugars, proteins, liquids, butyl rubbers, polyethylenes, polystyrenes, silicon polymers, and others. SEC has been mainly employed to study complex, biochemical, or highly polymerized molecules.

The importantance of SEC becomes apparent in the separation and characterization of molecules of different molecular weights. Molecules of similar molecular weights could be separated by a proper selection of the appropriate gel and column length. Desalting is the process of separating large molecules of biological origin from ionizable and inorganic species. For example, hemoglobin can be separated from sodium chloride by employing a column of Sephadex gel.

SEC can be also be used for preparative purpose. However, the main applications of SEC are as follows:

1. *Purification*: Biological macromolecules such as proteins, viruses, enzymes, nucleic acids, and hormones. Polysaccharides can be separated and purified by the correct selection of gel matrix.
2. *Determination of molecular weight*: Molecular weight is the important parameter in determining the effluent volumes of globular proteins. In a certain range of molecular weight, it is a linear function of the logarithm of effluent volume.
3. *Solution concentration*: A solution of high molecular weight can be concentrated by the use of dry sephadex G-25.
4. *Desalting*: High-molecular-weight compounds can also be desalted by using a column of sephadex G-25.
5. *Protein-building studies*: The reversible binding of a ligand to macromolecules can also be studied using exclusion chromatography. To do so, a sample of protein mixture, equilibrated with ligand solution, is coated on to a column of a suitable gel (e.g., G-25).

6.6 ION EXCHANGE CHROMATOGRAPHY

6.6.1 Introduction

Ion chromatography has evolved over many years. Two researchers, Spedding and Powell, worked on separating rare earth metals using displacement ion-exchange chromatography and published several research outcomes in the beginning of 1947 (Fritz, 2004). Later they worked on applying ion exchange chromatography for the separation of 14N and 15N isotopes in ammonia. Kraus and Nelson worked on developing a separation method for metal ions. Many metal ions such as chloride, nitrate, and fluoride complexes were separated using anion chromatography, and the results were published in the beginning of 1950 (Gjerde et al., 1979). Modern ion chromatography uses innovative separation methods developed by Small, Bauman, and Stevens. The method of suppressed conductivity detection is widely applied for quick separation of anions and cations. Gjerde et al. (1979) published a method for anion and cation chromatography using nonsuppressed conductivity detection.

6.6.2 Principle

Ion exchange chromatography was developed especially for the separation of molecules based on the differences in the charge. Biological macromolecules such as proteins, enzymes, amino acids, and nucleic acids can be purified with this technique (Lucy, 2003). Similar to other chromatographic techniques, ion exchange chromatography also includes both an aqueous mobile phase and an inert organic stationary phase matrix, which possess ionizable functional groups of countercharge. The two forms of ion exchange chromatography such as anion and cation are developed because these counterions exist in a state of equilibrium between the mobile and stationary phases (Figure 6.23). Exchangeable matrix counterions include acids (COO^-), single charged monoatomic ions (Na^+, K^+, Cl^-), hydroxide groups (OH), and protons (H^+).

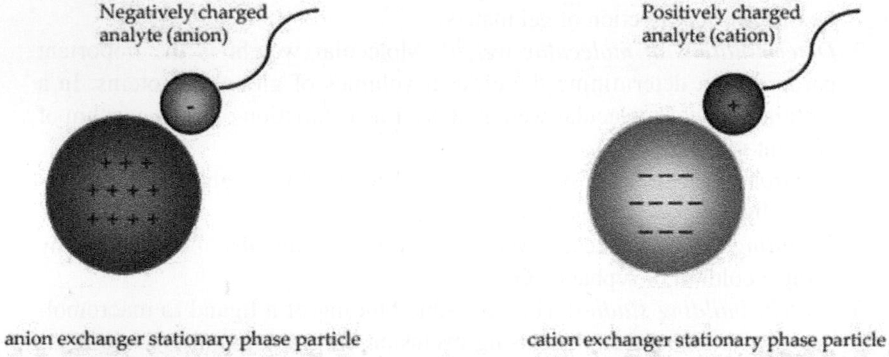

FIGURE 6.23 A schematic diagram of ion exchange chromatography.

Biomolecules are charged with a net positive or a net negative charge by applying the pH-dependent ionization of electrolyte groups (weak acids or bases). The mobile pH determines the net charge present on both the functional groups on the stationary phase matrix and on individual proteins. Some proteins carry both the net positive charge attributed to lysine and arginine side chains and the net negative charge attributed to glutamate and aspartate side chains, and these proteins are termed polyampholyte (Polgár, 2002). Generally, proteins are negatively charged above their isoelectric point (pI) (that is, pH at which a protein has zero net charge), and vice versa. Thus, each protein has a unique net charge on it at a given pH based on its pI value. Hence, mobile phase pH is used as a tool for finding the net charge on a protein of interest within a mixture and also to ensure whether the charge is opposite to that of the matrix functional group. This results in binding to the matrix (adsorption) by displacing the counterion functional group. The contaminants having opposite charge are not replaced with counterions. Finally, the bound protein analytes are eluted by using two properties: (1) pH and (2) ionic strength. Changes in the mobile phase pH helps in altering the net charge of the bound protein and the matrix-binding capacity. It means that, by raising the amount of similarly charged ions in the mobile phase, it competes with and displaces the bound ionic species. For example, during anion-exchange chromatography, the addition of negatively charged chloride ions competitively displaces the negatively charged protein molecules. The proteins can also be displaced and eluted by gradually increasing the amount of salt in the mobile phase, which results in an increase in the affinity between the salt ions and the functional groups and a decrease in the affinity between protein and functional groups (Männisto et al., 2007).

6.6.2.1 Choice of Buffers

1. Cationic buffers are used in anionic exchange chromatography.
2. Anionic buffers are used in cationic exchange chromatography.
3. The buffer should be selected so that it can withstand changes in pH in the system. This can be done by keeping the p*K* value of the buffer close to the pH of the system (Table 6.4) (Gjerde et al., 1979).

TABLE 6.4
Buffers with Different pH for Ion Exchange Chromatography

Buffer	pH Range
Ammonium formate	3–5
Pyridinium formate	3–6
Pyridinium acetate	4–6
Ammonium acetate	4–6
Ammonium carbonate	8–10

6.6.3 Application

Ion exchange chromatography is a widely applied purification technique because of its major advantages, such as wide applicability, large sample-handling capacity, high resolving ability, simplicity in scaling-up, and automation (Knudsen et al., 2001).

1. In the pharmaceutical industry, ion exchange chromatography is applied to determine the purity and contents of a substance, and to control drug production.
2. In semiconductor production, it is used for finding trace amounts of ions to test for organic solvent purity (Bonn, 1987).
3. In the cement industry, it is used to determine the presence of heavy metals.
4. In the chemical industry, it is used to determine the purity and the presence of heavy metals.
5. In the petrochemical industry, it is used to determine the presence of amines, mineral acids, and sulfur ions in solutions.
6. In the metallurgical industry, it is used to determine the composition of steel and other alloys.
7. In the paper industry, it is used to determine the presence of chlorine, transition metals, and sulfur in wastewater.

6.7 INVERSE GAS CHROMATOGRAPHY

6.7.1 Introduction

Inverse gas chromatography (IGC) is a highly versatile, sensitive, and fast technique for studying the properties of substances. It was developed as a gas phase technique to investigate the surface and bulk properties of materials like particulate and fibrous. Because of the wide applicability of IGC in determining the physicochemical properties of different forms of solids (films, fibers, and crystalline and amorphous powders), it became a powerful technique in investigating materials. The principle of IGC is similar to that of analytical gas chromatography (GC), which makes it easier for the scientist to understand and work. The basic difference in principle between IGC and GC is the role of two phases: mobile and stationary. In IGC, a known vapor is used as a probe for investigating the surface information of the substance placed in the chromatographic column. So far, IGC studies involve the use of semiautomated or manual equipment that is built at labs. Because of this, IGC poses certain drawbacks in the reproducibility of experimental conditions (Mohammadi-Jam and Waters, 2014).

6.7.2 Principle

Chromatographic analysis was first developed as a direct method, allowing the separation, quantification, and identification of the components of a complex mix (James and Martin, 1952). Soon the same authors proposed using this technique to determine the thermodynamic properties of a melted polymer impregnated on a finely divided solid (powder). This is the founding principle of IGC.

In order to perform such determinations, the role of the mobile and stationary phases are reversed. Here it is the stationary phase that is analyzed, and a column is filled with the material to be characterized. The experiment procedure involves the series of vapor injections or pulses, and thus it travels through the column packed with the target sample. Detectors like flame ionization detectors (FIDs) can be used to determine the retention time of the vapor. The bulk and surface property of the target sample can be studied by changing the column conditions, such as vapor probe molecule, flow rate, or temperature (Ho and Heng, 2013). Figure 6.24 shows IGC and analytical gas chromatography.

6.7.2.1 The Filling of the Chromatographic Column

The first step, the first critical point, of an IGC analysis is the filling of the column with the solid of interest. The technique of filling columns with ideal powders having a monodisperse granular size, that is, similar to the standard chromatographic stationary phase like Chromosorb®, is well known. But IGC often deals with much divided solids, having a mean particle diameter of a few microns (fillers for polymers like talc, silica, or carbon black) or with fibers (textile fibers or hairs, for example). Of course, each type of solid is a particular case, requiring an adapted filling protocol (Schreiber and Lloyd, 1989).

6.7.2.2 The Inverse Chromatography Techniques

The inverse chromatography techniques can be divided into two groups depending on the applied mobile phase. Indeed, the latter can be liquid or gaseous. If the mobile phase is liquid (solvent), the technique is called inverse liquid chromatography (ILC). When using a gaseous carrier phase, the technique is called inverse

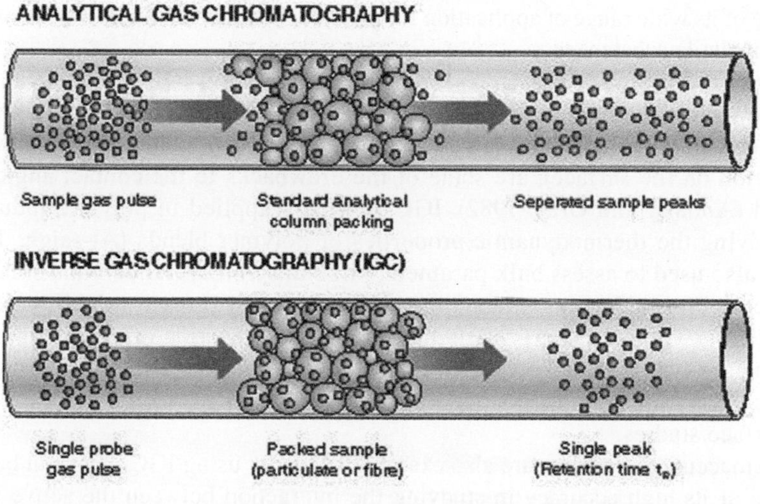

FIGURE 6.24 Sketch of typical gas chromatography (GC) and inverse gas chromatography (IGC).

gas chromatography (IGC). The ILC or IGC measurements involve the injection of known compounds as molecular probes to obtain chromatograms. Obviously, the measured retention times, as well as the shape of the obtained peaks, are based on the interactions occurring between the molecular probes and the investigated material (stationary phase). Therefore, the exploitation of measured retention times and/or of the shape of the chromatograms gives access to numerous physicochemical properties of the analyzed materials (Nastasović and Onjia, 2008). The IGC is the most common inverse chromatography technique. IGC is classified into two methods based on the molecular probe amounts injected:

1. *The IGC at infinite dilution (IGC-ID) method*: This method deals with the injection of a very low volume of molecular probes. In this case, interactions between molecular probes are insignificant, and the retention time is based on the interaction within the stationary phase. IGC-ID is employed to study the physicochemical properties of solid surfaces but also to find transition temperatures and to characterize liquids (for example, molten polymers) for the analysis of solubility parameters.
2. *The IGC at finite concentration (IGC-FC) method*: This method involves the injection of a finite amount of molecular probe. The surface coverage ratio close to monolayer can be reached. This method is used to determine properties such as desorption isotherms, surface heterogeneity, and specific surface area.

6.7.3 Application

Inverse chromatography techniques are potential investigation tools for research and development (R&D) departments and quality control. Numerous physicochemical properties can be measured with IGC. IGC has become the most powerful technique because of its wide range of application for physicochemical characterization of various nonvolatile substances.

Polymers are the most often researched material using IGC. Polymer surfaces are characterized using angle measurements of IGC. Surface heterogeneity and roughness, and also absorption of the liquid into the polymer bulk (instead of adsorption on the surface), are some of the drawbacks to the contact angle IGC method (Anhang and Gray, 1982). IGC is widely applied in polymer industries for studying the thermodynamic properties of polymer blends (Al-saigh, 1997). IGC is also used to assess bulk parameters like miscibility, Hildebrand solubility parameter, Flory–Huggins interaction parameters, and Hansen solubility parameters (HSPs). Guillet and Al-Saigh discussed the application of this technique in studying the synthetic and natural polymers. They investigated various properties such as glass transition, diffusion, degree of crystallinity, rate of crystallization, and surface studies.

Pharmaceutical powders are also examined broadly using IGC. IGC can be used because of its high accuracy in studying the interaction between the active ingredients in a product. The heterogeneity in surface energy of pharmaceutical powders plays an important role in analyzing product quality. The determination of free energy by the contact angle method using conventional methods may lead to erroneous results because of interference by the experimental conditions (Grimsey et al., 2002). Therefore, study of surface characterization of drug delivery systems (DDSs) such as active pharmaceutical ingredients (APIs) and excipients by IGC has become more popular. It is also used in estimating storage conditions and shelf life of pharmaceutical products using properties like crystallization rates (Guo et al., 2013). Pharmaceutical products are examined under various conditions such as temperature, ambient moisture, and compaction pressure.

6.8 GRADIENT ELUTION

A gradient method is employed when the samples are not easily separated by isocratic methods because of their wide *k* (capacity factor) range. The strength of the eluent can be increased by gradually increasing the composition of the mobile phase during the separation. This results in a reduction in analysis time and an improvement in the quality of the separation as well as the detection limit. The easiest gradient method involves the use of binary linear gradients. Gradient elution has some advantages over isocratic elution, although it is difficult to do.

Complex chromatograms with widely and irregularly spaced retentions were obtained with many compounds of differing polarities. The total analysis time is unacceptably long with isocratic methods. In terms of retention and selectivity:

1. Some leave the column more or less unretained ($k < 1$).
2. Some leave the column in a workable time ($1 < k < 10$).
3. Some are strongly retained ($k \gg 10$).
4. Some are not totally separated from each other (inadequate selectivity or resolution).

The unretained compounds are combined together and stay hidden under the solvent peak in the chromatogram. The determination of retention time and peak area is the difficult part in chromatography. In a few cases, the peaks are broadened or split, and so they do not provide the correct peak area. Sometimes, peak broadening is too extreme and it results in the incorrect determination of compounds (Chamarthy, 2007).

The gradient elution is recommended for: (1) samples with a wide range of *k*, (2) samples containing large molecules >1000 in molecular weight, and (3) samples containing late eluting interferences that can either foul the column or overlap subsequent chromatograms.

6.9 TROUBLESHOOTING CHROMATOGRAPHY

First, a model method is set up before the system develops an error. The method should be 100% reliable. The set of data for pressure, repeatability, noise, and so on, is determined using this method. Second, always follow the procedure of operation and maintenance. Third, the problem should be clearly identified by comparing the erroneous chromatogram with a known accurate one, with complete details such as retention times, peak height, elution order, T_0 peak. The tips to avoid error in HPLC equipment are summarized below:

1. Check everything without any assumptions.
2. Ensure what is known initially.
3. Test one thing at a time.
4. Maintain notes on the procedure followed.
5. Use a representative method of known conditions to discover the fault.
6. Decreasing mobile phase pH results in error chromatogram such as peak tailing and reduction in silanols interactions.
7. Peak broadening can be caused by improper HPLC system connections.
8. Acid wash can improve peak shape.
9. Employ columns designed for the chosen pH.
10. Solvent and mobile phase should be selected properly to avoid the peak shape problems such as peak splitting or broadening.
11. The peak distortion can be due to sample overload. Reduce the sample size.

6.10 CONCLUSION

Chromatography plays a vital role in many industries such as pharmaceutical, biochemical, environmental, and food. Many methods are being developed to couple the chromatography technique with other identification techniques, such as mass spectrometry (LC-MS, GC-MS) and nuclear magnetic resonance spectroscopy (LC-NMR) to help scientists and technicians understand the complex mixtures by separation and analysis. Recent advancements in chromatography are aimed at improving the level of automation in order to develop a laboratory of the future, all in an automated way.

REFERENCES

Al-saigh, Z. Y. 1997. Review: Inverse gas chromatography for the characterization of polymer blends. *International Journal of Polymer Analysis and Characterization* 3 (3): 249–291.

Anhang, J., and D. G. Gray. 1982. Surface characterization of Poly(Ethylene Terephthalate) film by inverse gas chromatography. *Journal of Applied Polymer Science* 27 (1): 71–78.

Arsenis, C., and D. B. Mccormick. 1964. Purification of liver flavokinase by column chromatography on flavin-cellulose compounds. *The Journal of Biological Chemistry* 239: 3093–3097.

Axén, R., J. Porath, and S. Ernback. 1967. Chemical coupling of peptides and proteins to polysaccharides by means of cyanogen halides. *Nature* 214 (5095): 1302–1304.

Bautz, E. K., and B. D. Hall. 1962. The isolation of T4-specific RNA on a DNA-cellulose column. *Proceedings of the National Academy of Sciences of the United States of America* 48: 400–408.

Bio-Rad. 2019. Affinity chromatography | LSR | Bio-Rad. Accessed January 4, 2019. http://www.bio-rad.com/en-us/applications-technologies/affinity-chromatography?ID=LUSMJIDN.

Bird, I. M. 1989. High performance liquid chromatography: Principles and clinical applications. *British Medical Journal* (*Clinical Research Ed.*) 299 (6702): 783–787.

Bonn, G. 1987. High-performance liquid chromatographic isolation of ^{14}C-labelled gluco-oligosaccharides, monosaccharides and sugar degradation products on ion-exchange resins. *Journal of Chromatography A* 387: 393–398.

Bose, A., T. W. Wong, and N. Singh. 2013. Formulation development and optimization of sustained release matrix tablet of itopride HCl by response surface methodology and its evaluation of release kinetics. *Saudi Pharmaceutical Journal: SPJ* 21 (2): 201–213.

Caballero, B., P. M. Finglas, and F. Toldrá. 2015. *Encyclopedia of Food and Health.* Oxford, UK: Elsevier Science.

Campbell, D. H., E. Luescher, and L. S. Lerman. 1951. Immunologic adsorbents: I. Isolation of antibody by means of a cellulose-protein antigen. *Proceedings of the National Academy of Sciences of the United States of America* 37 (9): 575–578.

Chamarthy, S. P. 2007. The different roles of surface and bulk effects on the functionality of pharmaceutical materials. Theses and Dissertations Available from ProQuest, January.

Chen, J., and D. S. Hage. 2004. Quantitative analysis of allosteric drug-protein binding by biointeraction chromatography. *Nature Biotechnology* 22 (11): 1445–1448.

Clonis, Y. D., N. E. Labrou, V. P. Kotsira, C. Mazitsos, S. Melissis, and G. Gogolas. 2000. Biomimetic dyes as affinity chromatography tools in enzyme purification. *Journal of Chromatography A* 891 (1): 33–44.

Cuatrecasas, P., M. Wilchek, and C. B. Anfinsen. 1968. Selective enzyme purification by affinity chromatography. *Proceedings of the National Academy of Sciences of the United States of America* 61 (2): 636–643.

Das, S. K., and I. Mukherjee. 2011. Effect of light and PH on persistence of flubendiamide. *Bulletin of Environmental Contamination and Toxicology* 87 (3): 292–296.

Determann, H. 1969. Introduction. In *Gel Chromatography*, pp. 1–12. Berlin, Germany: Springer.

Dolan, J. W., and L. R. Snyder. 2012. Gradient elution chromatography. In *Encyclopedia of Analytical Chemistry.* Chichester, UK: John Wiley & Sons, Ltd.

Ettre, L. S. 1980. Evolution of liquid chromatography: A historical overview. Cs. Horvath, Ed., *High-Performance Liquid Chromatography: Advances and Perspectives*, Vol. 1. New York, Academic Press, pp. 1–74.

Ettre, L. S. 2000. Chromatography: The separation technique of the 20th century. *Chromatographia* 51 (1–2): 7–17.

Ettre, L. S., and J. V. Hinshaw. 2008. *Chapters in the Evolution of Chromatography.* London, UK: Published by Imperial College Press and distributed by World Scientific Publishing Co.

Ettre, L. S., and K. I. Sakodynskii. 1993. M. S. Tswett and the discovery of chromatography I: early work (1899–1903). *Chromatographia* 35 (3–4): 223–231.

Fritz, J. S. 2004. Early milestones in the development of ion-exchange chromatography: A personal account. *Journal of Chromatography A* 1039 (1–2): 3–12.

Gerberding, S. J., and C. H. Byers. 1998. Preparative ion-exchange chromatography of proteins from dairy whey. *Journal of Chromatography A* 808 (1–2): 141–151.

Gilroy, J. J., J. W. Dolan, and L. R. Snyder. 2003. Column selectivity in reversed-phase liquid chromatography: IV. Type-B Alkyl-Silica Columns. *Journal of Chromatography A* 1000 (1–2): 757–778.

Gjerde, D. T., J. S. Fritz, and G. Schmuckler. 1979. Anion chromatography with low-conductivity eluents. *Journal of Chromatography A* 186: 509–519.

Grimsey, I. M., J. C. Feeley, and P. York. 2002. Analysis of the surface energy of pharmaceutical powders by inverse gas chromatography. *Journal of Pharmaceutical Sciences* 91 (2): 571–583.

Guo, Y., E. Shalaev, and S. Smith. 2013. Physical stability of pharmaceutical formulations: Solid-state characterization of amorphous dispersions. *TrAC Trends in Analytical Chemistry* 49: 137–144.

Hage, D. S. 2002. High-performance affinity chromatography: A powerful tool for studying serum protein binding. *Journal of Chromatography B* 768 (1): 3–30.

Hage, D. S., J. Anguizola, O. Barnaby, A. Jackson, M. J. Yoo, E. Papastavros, E. Pfaunmiller, M. Sobansky, and Z. Tong. 2011. Characterization of drug interactions with serum proteins by using high-performance affinity chromatography. *Current Drug Metabolism* 12 (4): 313–328.

Hage, D. S., and J. Austin. 2000. High-performance affinity chromatography and immobilized serum albumin as probes for drug- and hormone-protein binding. *Journal of Chromatography B, Biomedical Sciences and Applications* 739 (1): 39–54.

Hawkes, S. J. 1983. Modernization of the van Deemter equation for chromatographic zone dispersion. *Journal of Chemical Education* 60 (5): 393.

Heftmann, E. 1983. *Chromatography: Fundamentals and Applications of Chromatographic and Electrophoretic Methods.* Amsterdam, the Netherlands: Elsevier Scientific Publishing Co.

Hermanson, G. T. 1992. *Immobilized Affinity Ligand Techniques/Greg T. Hermanson, A. Krishna Mallia, and Paul K. Smith.—Version Details—Trove*, 1st ed. San Diego, CA: Academic Press.

Ho, R., and J. Y. Y. Heng. 2013. A review of inverse gas chromatography and its development as a tool to characterize anisotropic surface properties of pharmaceutical solids. *KONA Powder and Particle Journal* 30: 164–180.

Hong, P., S. Koza, and E. S. P. Bouvier. 2012. Size-exclusion chromatography for the analysis of protein biotherapeutics and their aggregates. *Journal of Liquid Chromatography & Related Technologies* 35 (20): 2923–2950.

Horváth, C. 1988. *High-Performance Liquid Chromatography: Advances and Perspectives*, Vol. 5. San Diego, CA: Academic Press.

James, A. T., and A. J. P. Martin. 1952. Gas-liquid partition chromatography; the separation and micro-estimation of volatile fatty acids from formic acid to dodecanoic acid. *The Biochemical Journal* 50 (5): 679–690.

Karger, B. L., L. R. Snyder, and C. Horváth. 1973. *An Introduction to Separation Science.* London, UK: Wiley.

Knudsen, H. L., R. L. Fahrner, Y. Xu, L. A. Norling, and G. S. Blank. 2001. Membrane ion-exchange chromatography for process-scale antibody purification. *Journal of Chromatography A* 907 (1–2): 145–154.

Lowe, C. R., S. J. Burton, N. P. Burton, W. K. Alderton, J. M. Pitts, and J. A. Thomas. 1992. Designer dyes: 'Biomimetic' ligands for the purification of pharmaceutical proteins by affinity chromatography. *Trends in Biotechnology* 10: 442–448.

Lucy, C. A. 2003. Evolution of ion-exchange: From Moses to the Manhattan Project to modern times. *Journal of Chromatography A* 1000 (1–2): 711–724.

Majors, R. E. 2015. Historical developments in HPLC and UHPLC column technology: The past 25 years. LCGC North America 33 (11): 818–840. http://www.chromatographyonline.com/historical-developments-hplc-and-uhplc-column-technology-past-25-years.

Majors, R. E. 2019. Historical developments in HPLC and UHPLC column technology: The past 25 years. Accessed January 4, 2019.

Männisto, P. T., J. Vanäläinen, A. Jalkanen, and J. A. García-Horsman. 2007. Prolyl oligopeptidase: A potential target for the treatment of cognitive disorders. *Drug News & Perspectives* 20 (5): 293.

Martin, A. J. P. 1950. Partition chromatography. *Annual Review of Biochemistry* 19 (1): 517–542.

Mohammadi-Jam, S., and K. E. Waters. 2014. Inverse gas chromatography applications: A review. *Advances in Colloid and Interface Science* 212: 21–44.

Nastasović, A. B., and A. E. Onjia. 2008. Determination of glass temperature of polymers by inverse gas chromatography. *Journal of Chromatography A* 1195 (1–2): 1–15.

Ó'Fágáin, C., P. M. Cummins, and B. F. O'Connor. 2011. Gel-filtration chromatography. *Methods in Molecular Biology* 681: 25–33.
Ohzeki, K., T. Kambara, and K. Kodama. 1976. Effect of column length on HETP in gas chromatography. *Journal of Chromatography A* 121 (2): 199–204.
Patil, S. V., R. Y. Patil, and V. U. Barge. 1970. Size-exclusion chromatography in biotech industry. *Research & Reviews: Journal of Microbiology and Biotechnology* 3 (4): 11–14.
Pavia, D. L. 2005. *Introduction to Organic Laboratory Techniques: A Small Scale Approach.* Pacific Grove, CA: Thomson Brooks/Cole.
Petrova, O. E., and K. Sauer. 2017. High-Performance Liquid Chromatography (HPLC)-based detection and quantitation of cellular c-Di-GMP. *Methods in Molecular Biology (Clifton, NJ)* 1657: 33–43.
Phillips, T. M., W. D. Queen, N. S. More, and A. M. Thompson. 1985. Protein A-coated glass beads. Universal support medium for high-performance immunoaffinity chromatography. *Journal of Chromatography* 327: 213–219.
Polgár, L. 2002. The prolyl oligopeptidase family. *Cellular and Molecular Life Sciences: CMLS* 59 (2): 349–362.
Porath, J. 1997. From gel filtration to adsorptive size exclusion. *Journal of Protein Chemistry* 16 (5): 463–468.
Ray, N. H. 2007. Gas chromatography. II. The separation and analysis of gas mixtures by chromatographic methods. *Journal of Applied Chemistry* 4 (2): 82–85.
Regueiro, J., C. Sánchez-González, A. Vallverdú-Queralt, J. Simal-Gándara, R. Lamuela-Raventós, and M. Izquierdo-Pulido. 2014. Comprehensive identification of walnut polyphenols by liquid chromatography coupled to linear ion trap-orbitrap mass spectrometry. *Food Chemistry* 152: 340–348.
Rogatsky, E. 2016. Modern high performance liquid chromatography and HPLC 2016 International Symposium. *Journal of Chromatography & Separation Techniques* 7 (4): 1–1.
Roque, A. C. A., and C. R. Lowe. 2008. Affinity chromatography. In *Affinity Chromatography* edited by M. Zachariou, pp. 1–23. Totowa, NJ: Humana Press.
Ruhn, P. F., S. Garver, and D. S. Hage. 1994. Development of dihydrazide-activated silica supports for high-performance affinity chromatography. *Journal of Chromatography A* 669 (1–2): 9–19.
Schever, P. 1980. *Marine Natural Products V3: Chemical and Biological Perspectives.* New York: Elsevier Science.
Schreiber, H. P., and D. R. Lloyd. 1989. Overview of inverse gas chromatography. D. R. Lloyd, T. C. Ward, H. P. Schreiber, Eds., *Inverse Gas Chromatography: Characterization of Polymers and Other Materials*, pp. 1–10. Washington, DC: American Chemical Society.
Scott, R. P. W., and A. Weissberger. 1976. *Contemporary Liquid Chromatography.* New York: Wiley.
Sharon, N. 1998. Lectins: From obscurity into the limelight. *Protein Science: A Publication of the Protein Society* 7 (9): 2042–2048.
Skoog, D. A., F. J. Holler, and T. A. Nieman. 1998. *Principles of Instrumental Analysis.* Published in 1998 in Belmont (Calif.) by Brooks/Cole, 5th ed. Philadelphia, PA: Saunders College Publishing.
Stauffer, E., J. A. Dolan, and R. Newman. 2008. *Fire Debris Analysis*, 1st ed. Boston, MA: Academic Press.
Sun, P., C. Wang, Z. S Breitbach, Y. Zhang, and D. W. Armstrong. 2009. Development of new HPLC chiral stationary phases based on native and derivatized cyclofructans. *Analytical Chemistry* 81 (24): 10215–10226.
Telepchak, M. J., G. Chaney, and T. F. August. 2004. *Forensic and Clinical Applications of Solid Phase Extraction.* Totowa, NJ: Humana Press.

Thammana, M. 2016. A review on High Performance Liquid Chromatography (HPLC). *Research & Reviews: Journal of Pharmaceutical Analysis* 5 (2): 1–7.

Tu, C., P. A. Rudnick, M. Y. Martinez, K. L. Cheek, S. E. Stein, R. J. C. Slebos, and D. C. Liebler. 2010. Depletion of abundant plasma proteins and limitations of plasma proteomics. *Journal of Proteome Research* 9 (10): 4982–4991.

Vallverdú-Queralt, A., O. Jáuregui, A. Medina-Remón, C. Andrés-Lacueva, and R. M. Lamuela-Raventós. 2010. Improved characterization of tomato polyphenols using liquid chromatography/electrospray ionization linear ion trap quadrupole orbitrap mass spectrometry and liquid chromatography/electrospray ionization tandem mass spectrometry. *Rapid Communications in Mass Spectrometry* 24 (20): 2986–2992.

Walls, D., R. McGrath, and S. T. Loughran. 2011. A digest of protein purification. *Methods in Molecular Biology (Clifton, N.J.)* 681: 3–23.

Wang, G., J. De, J. S. Schoeniger, D. C. Roe, and R. G. Carbonell. 2004. A hexamer peptide ligand that binds selectively to staphylococcal enterotoxin B: Isolation from a solid phase combinatorial library. *Journal of Peptide Research* 64 (2): 51–64.

Wang, T., D. Li, B. Yu, and J. Qi. 2017. Screening inhibitors of xanthine oxidase from natural products using enzyme immobilized magnetic beads by high-performance liquid chromatography coupled with tandem mass spectrometry. *Journal of Separation Science* 40 (9): 1877–1886.

Wang, X., D. R. Stoll, P. W. Carr, and P. J. Schoenmakers. 2006. A graphical method for understanding the kinetics of peak capacity production in gradient elution liquid chromatography. *Journal of Chromatography A* 1125 (2): 177–181.

Warren, W. J., and G. Vella. 1995. Principles and methods for the analysis and purification of synthetic deoxyribonucleotides by high-performance liquid chromatography. *Molecular Biotechnology* 4 (2): 179–199.

Werner, A., W. Siems, U. Kuckelkorn, T. Grune, and C. Schreiter. 1990. Anion-exchange and size-exclusion chromatography of proteins in mouse ascites fluid: Changes in the protein pattern during tumour growthIn vivo. *Chromatographia* 29 (7–8): 351–354.

Zachariou, M. 2008. Affinity chromatography: Methods and protocols. Preface. *Methods in Molecular Biology (Clifton, N.J.)* 421: vii–viii.

Zotou, A. 2012. An overview of recent advances in HPLC instrumentation. *Open Chemistry* 10 (3): 554–569.

7 Biosorption

Billie Yan Zhang Hiew, Lai Yee Lee, Suchithra Thangalazhy-Gopakumar, and Suyin Gan

7.1 INTRODUCTION

The generation of toxic contaminants from anthropogenic development and rapid industrialization has caused severe water pollution, leading to global concerns about access to clean drinkable water. Even in trace amounts, water pollutants such as synthetic dyes, heavy metal ions, radionuclides and micropollutants have been reported to have been detrimental effects on ecosystems and human health (Vijayaraghavan and Balasubramanian, 2015; Hiew et al., 2018). There are two sources of water pollution: direct and indirect. An example of a direct source is the direct discharge of waste effluent from process plants; examples of indirect sources are the contaminants that enter the water supply via soil, sewage systems and rainwater (Vijayaraghavan and Yun, 2008).

To address the aforementioned issue, various industrial treatment technologies were developed to prevent and limit the toxic pollutants to a safe concentration level and therefore minimize the harmful effects. Conventional treatment methods such as chemical precipitation, flocculation-coagulation, ionic exchange, membrane filtration and reverse osmosis have demonstrated different degrees of treatment efficiency (Gupta et al., 2018), and these methods have advantages and drawbacks. Major drawbacks are the generation of secondary waste and high operating costs as well as chemical requirements which complicate the water purification process. Recently, bioremediation methods such as bioaccumulation, biosorption and phytoremediation received growing attention in wastewater treatment processes as they exhibit high removal efficiency, are environmentally benign and low cost, and do not generate secondary pollutants. Among them, biosorption is a potential bioremediation method for future water purification.

Biosorption is defined as the passive and metabolically independent uptake of pollutant by dead or biologically inactive materials through different physicochemical mechanisms (Gadd, 2009; Vijayaraghavan and Balasubramanian, 2015). Figure 7.1 illustrates the interaction between the solid phase (biosorbent) and the pollutant (biosorbate) in the liquid phase through typical sorption process.

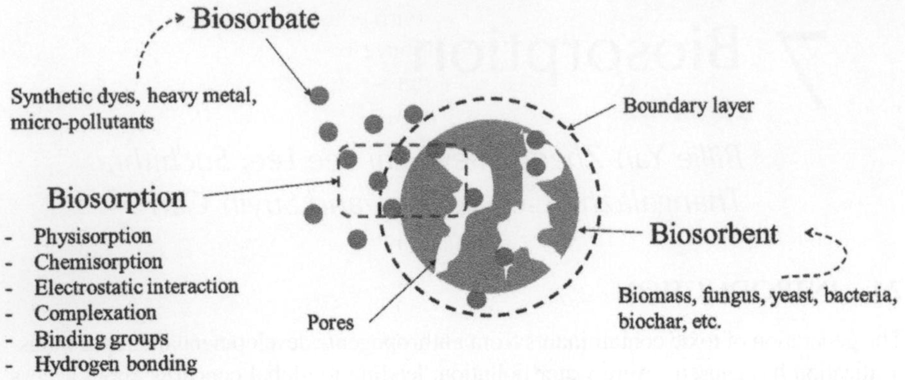

FIGURE 7.1 Biosorption process.

The frequently reported biosorbents for wastewater decontamination include fungi, bacteria, algae, agricultural wastes, polysaccharide materials and industrial wastes. They exhibit good removal efficiency toward water pollutants (Lu et al., 2012; Bouras et al., 2017; Li et al., 2018; Ngabura et al., 2018).

7.2 SELECTION OF BIOSORBENT

Any kind of biological material has a degree of affinity toward inorganic and organic pollutants, thus there is a wide range of potential biosorbents that can be utilized in the biosorption process. During the screening of potential biosorbents, care and consideration should be equally given to the biosorption capacity, cost effectiveness, availability and mechanical strength to ensure feasible biosorption operation. The criteria for selecting a suitable biosorbent for wastewater decontamination are as follows (Vijayaraghavan and Balasubramanian, 2015):

- No secondary pollutants generated
- Cost effective
- High regenerability and reusability
- No requirement for chemical modification
- No pre-treatment necessity
- High biosorption capacity for pollutants
- Good stability for operation at a wide range of pH, temperature and other process parameters
- High adaptation into different system designs

Even though these criteria are highly favorable, different types of biosorbents possess different natural properties that may not meet all the criteria. Therefore, biosorbents can be technically tailored to suit the biosorption application via modification.

Cost effectiveness is the main factor in biosorbent selection to ensure a feasible biosorption process. The origin of the biomass also has to be considered during the selection of the biosorbent to secure a continuous supply for large-scale operation. Currently, a wide range of biomasses have been investigated for biosorbent preparation, and these include microbial biomass (bacteria, archaea, fungi, yeast and microalgae), seaweed (macroalgae), industrial wastes (food and fermentation wastes, anaerobic and activated sludges), agricultural wastes (fruit and vegetable wastes) and others.

Biosorbents derived from bacterial, fungal and yeast exhibited excellent biosorption capacity due to their cell wall composition which facilitated the biosorption process (Vijayaraghavan and Yun, 2008; Akar et al., 2013). However, most of these biosorbents were first cultivated prior to use in biosorption. The cultivation process increased the overall process cost as well as the uncertainty in the supply of biomass (Vijayaraghavan and Balasubramanian, 2015). Microbial wastes generated from industrial processes (food, fermentation and pharmaceutical) can also be used as effective biosorbents. These microbial wastes counter the waste disposal issue as well as generate revenue for the industries.

Marine algae (seaweeds) is another potential biosorbent. They are abundant and fast growing. Typically, marine algae represent a threat to the tourism industry because they damage the local marine environment and foul beaches (Volesky, 2001). Hence, it is beneficial to a local community that relies on tourism to utilize marine algae in biosorption process. However, it should be noted that marine algae are not classified as waste, and they are the sole source for the production of alginate, carrageenan and agar derivatives (Vijayaraghavan and Balasubramanian, 2015). Therefore, the selection of marine algae for biosorption must be done with care.

Industrial and agricultural wastes have been studied extensively as potential biosorbents in wastewater treatment because these biomasses are relatively inexpensive, easy to acquire and available for sustainable biosorbent production. Biosorbents developed from such biomass are reported to possess excellent biosorption capacity and reasonable rigidity (Zhang et al., 2016; Lee et al., 2017; Ungureanu et al., 2017; Moscatello et al., 2018).

In some cases, biomass requires modification to improve the biosorption performance. The treatment methods are discussed in the following section.

7.3 MODIFICATION OF BIOSORBENT

Different methods to manipulate the physicochemical properties of biomass for biosorption process are displayed in Figure 7.2.

Freely suspended biosorbents tend to cause particle agglomeration which leads to poor mechanical strength as well as biosorbent swelling. Therefore, solid-liquid separation and regeneration would be less efficient in large-scale biosorption operations (Fomina and Gadd, 2014). Immobilization techniques offer good mechanical strength via inert support, entrapment within polymer matrices and cross-linking mechanisms. By immobilizing the fragile biomass with proper support, the developed material can overcome the challenges in the recovery process (Wen et al., 2018). Magnetism is another modification method to recover freely suspended biosorbents through magnetic separation. This method can be achieved by doping Fe_2O_3

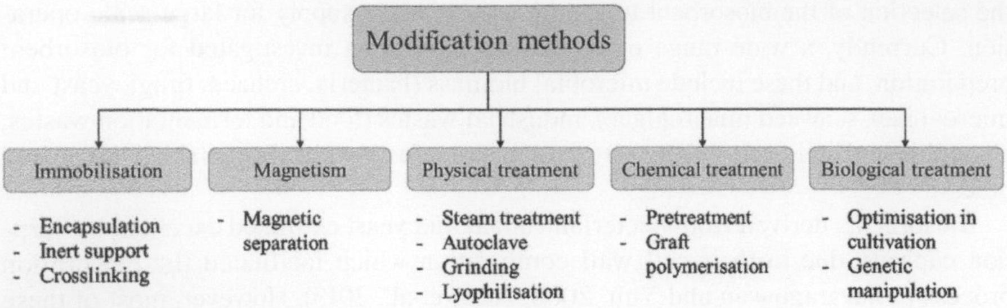

FIGURE 7.2 Common modification methods used in biosorbent development.

nanoparticles on the biomass surface (Zhang et al., 2016; Daneshfozoun et al., 2017). Immobilization and magnetism techniques can be applied to biosorbents like bacteria, yeast, microalgae, agricultural waste, and microorganisms to assist in easy solid-liquid separation for regeneration processes.

In addition, physical and chemical modifications are widely used to improve the biosorption performance of biomass. Physical modification such as autoclave, steam treatment, pyrolysis, lyophilization and particle reduction are used (Fomina and Gadd, 2014; Manna et al., 2015). In comparison, chemical modification such as pre-treatment with acid, alkali, organic solvent and polymer grafting improved the biosorption capacity significantly through the introduction of new functional groups or enrichment in specific functional groups (Kumar et al., 2017). Adsorbents such as biochar, activated carbon, acid- or alkali-treated biosorbent, oxidized biosorbent and chemically treated biosorbent are some of the commonly reported biosorbents modified via physical and chemical modification. The combination of biosorbent with natural clay material is another alternative to increase the mechanical properties of biosorbent as well as improve biosorption capacity. Commonly used clays in the modification of biomass are kaolinite, bentonite, sepiolite and montmorillonite. The effectiveness of biosorbent can also be improved through biological modification such as optimization in cell cultivation process and genetic manipulation. For instance, washing pre-treatment, surface functionalization and genetic engineering (metallothionein, phytochelatin, short peptide and metalloregulatory proteins) are commonly used methods for biological modification.

Despite the great improvement in biosorption performance via biosorbent modification, these modification processes often increase the overall production cost which offsets the major advantage of biosorption, which is low cost. In addition, several environmental, health and safety concerns are raised with regard to these modification methods, for instance, the harsh and hazardous chemicals used in physical and chemical modifications, the high dispersive energy radiation in graft polymerization, and the potential biological hazard and controversial genetically modified microorganism in biological modification (Fomina and Gadd, 2014). Hence, the nature of the biosorbent and the necessity of the modification process must be assessed to develop a low-cost, green and effective material for the biosorption process.

7.4 INFLUENCES ON BIOSORPTION PERFORMANCE

Biosorption performance is affected by the selection of biosorbent and by the operational parameters such as solution pH, process temperature, initial biosorbate concentration and contact time. Among the studied parameters, solution pH is the most important because it significantly influences the biosorbent characteristics, the activity of the surface functional groups and the solution chemistry of the biosorbate. Point of zero charge (pH_{PZC}) offers information on the surface charge on adsorbent based on the solution pH. In general, the adsorbent is positively charged when $pH < pH_{PZC}$, and negatively charged when $pH > pH_{PZC}$. In the case of dye biosorption, the operational pH range depends on the classification of dyes. Generally, biosorption of anionic dyes are favorable under acidic conditions, or when $pH < pH_{PZC}$, whereas biosorption of cationic dyes are favorable at neutral, alkaline condition, or when $pH > pH_{PZC}$ (Yagub et al., 2014; Ooi et al., 2017). In heavy metal biosorption, the solution pH has a dominant effect on the metal speciation and biosorption capacity. Overall, biosorption must be operated at optimum pH to enhance the biosorption capacity of selected biosorbent and minimize the structural damage to the biosorbent at extreme pH levels.

Biosorbent dosage is another strong factor affecting biosorption performance. In general, an increase in biomass dosage increases the amount of biosorbate uptake due to the increased availability of binding sites and specific surface area. However, under high biosorbent loading, particle agglomeration may occur which reduces the available surface area for biosorption and also extends the diffusional pathway of solute onto biosorbent (Ngabura et al., 2018). Therefore, the quantity of biosorbed solute per unit mass of biosorbent (biosorption capacity) decreases with increasing biosorbent dosage.

Biosorbent particle size plays a critical role in biosorption. Fine particles (<0.075 mm) generally favor the biosorption process because they offer greater surface area for pollutant uptake in a short residence time (Saha et al., 2017). However, the use of fine particle sizes reduces hydraulic conductivity which may increase the pressure drop in adsorption column. Apart from that, the size of the biosorbent used in the biosorption process should be mechanically stable and resilient to withstand the process pressures and other extreme operating conditions during regeneration. Hence, preliminary study on the effect of particle size is required to determine the suitable particle size of biosorbent to deliver a feasible biosorption process. Additional efforts such as granulation and immobilization can be applied on powdered biosorbent to enhance the mechanical properties of the selected biosorbent in a continuous process (Zhang et al., 2016).

Initial biosorbate concentration is reported to influence the biosorption process. At low initial biosorbate concentration, the ratio of the initial molar biosorbate species to the available binding sites is low. Therefore, fractional uptakes of biosorbate species do not depend on the initial biosorbate concentration. In contrast, at higher biosorbate concentrations, the biosorption uptake depends greatly on the initial concentration because there is a limited number of binding sites available (Xu et al., 2018). Hence, it is necessary to identify the potential maximum saturation capacity of a biosorbent by fitting the experimental results with theoretical models to design a suitable biosorption system.

The external film diffusion can be one of the rate-limiting steps in biosorption process. For batch biosorption system with suitable agitation speed, the mass transfer resistance can be minimized. The enhanced turbulence reduces the boundary layer thickness which in turn increases the diffusion rate of biosorbate from bulk liquid to the solid surface (Long et al., 2018).

Process temperature also affects the biosorption performance. At high temperatures (>50°C), biosorption is usually enhanced due to the increase in surface activity and kinetic energy of the biosorbate species; however, structural changes on the biosorbent can be anticipated when operating at high temperature (Tka et al., 2018). For exothermic biosorption processes, it was observed that the biosorption capacity of the biosorbent decreased at elevated temperatures. The type of biosorption process (exothermic or endothermic) can be determined by calculating the thermodynamic properties with experimental data. The biosorption is considered exothermic if the enthalpy change is negative or endothermic if the enthalpy change is positive.

7.5 BIOSORPTION THEORETICAL MODELS

The performance of biosorption processes can be further studied by fitting the experimental data with theoretical models. These models provide the equilibrium relationship between the biosorbent and adsorbate, as well as the rate and mechanism of biosorption.

7.5.1 Isotherm Models

Biosorption capacity (q_e) is often selected as the quantitative benchmark to describe and evaluate process performance. Various mathematical models have been established to describe the adsorption equilibrium. The commonly used equilibrium models are Langmuir, Freundlich, Temkin and Dubinin–Radushkevich (DR) models, as shown below:

Langmuir:

$$q_e = \frac{q_m K_L C_e}{1 + K_L C_e} \tag{7.1}$$

Hall's separation factor:

$$R_L = \frac{1}{1 + K_L C_O{}'} \tag{7.2}$$

Freundlich:

$$q_e = K_f C_e{}^{1/n} \tag{7.3}$$

Temkin:

$$q_e = \frac{RT}{b_T} \ln\left(A_T C_e\right) \tag{7.4}$$

DR:

$$q_e = q_{DR} e^{-K_{DR}\varepsilon^2} \tag{7.5}$$

Mean free sorption energy:

$$E = \frac{1}{\sqrt{2K_{DR}}} \tag{7.6}$$

where q_e (mg/g) is the equilibrium adsorption capacity, q_m (mg/g) is the Langmuir maximum adsorption capacity, K_L (L/mg) is the Langmuir adsorption constant, K_f ([mg/g][L/mg]$^{1/n}$) is the Freundlich constant, n is the Freundlich exponent, R is the universal gas constant (8.314 J/mol K), T is the temperature (K), A_T is the Temkin isotherm equilibrium binding constant (L/g), b_T is the Temkin isotherm constant, q_{DR} (mg/g) is the DR maximum sorption capacity, K_{DR} is the DR constant related to sorption energy, ε (=$RT \ln [1/1+C_e]$) is the Polanyi potential, C_o' (mg/L) is the highest initial concentration, and C_e (mg/L) is the equilibrium concentration.

The Langmuir model for biosorption assumes that the adsorbent surface has a finite number of homogeneous binding sites with equivalent binding energy and no interaction between adsorbates attached on adjacent sites (Langmuir, 1918). The Freundlich model assumes that multiple reversible adsorbates are adsorbed onto the heterogeneous adsorbent surface with logarithmic depletion in sorption energy as the surface coverage increases (Freundlich, 1906). The process favorability can be determined by Hall's separation factor (R_L) and the Freundlich exponent (n), as shown in Table 7.1 (Hall et al., 1966).

TABLE 7.1
Relationship between Isotherm Parameters and Isotherm Shape

Freundlich Exponent	Hall's Separation Factor	Description	Isotherm Shapes
$n = 0$	$R_L = 0$	Irreversible	Horizontal
$n < 1$	$R_L < 1$	Favorable	Concave
$n = 1$	$R_L = 1$	Linear	Linear
$n > 1$	$R_L > 1$	Unfavorable	Convex

Source: Tran, H.N. et al., *Water Res.*, 120, 88–116, 2017.

The Temkin model predicts the linear depletion of the sorption energy with surface coverage by neglecting extremely low and high adsorbate concentration (Temkin and Phyzev, 1940). The DR model is one of the temperature-dependent equilibrium models that represent an equilibrium isotherm with a high degree of rectangularity which is attributed to the porous structure of the adsorbent (Dubinin and Radushkevich 1947). Likewise, the K_{DR} value can be used to estimate the mean free sorption energy of the biosorption system and to determine if the system is governed by physisorption ($10 < E < 40$ kJ/mol) or chemisorption ($E = 40$–400 kJ/mol).

7.5.2 Kinetic Models

The kinetic of biosorption is another key element in designing a practical biosorption system. The kinetic profile determines the time period for the system to reach equilibrium and rate-limiting steps as well as the mechanism for biosorption. The instantaneous adsorption capacity at a specific time (q_t) is often used to study the biosorption kinetic. The biosorption kinetic can be determined by analyzing experimental data with empirical models such as pseudo-first-order (PFO) kinetic, pseudo-second-order (PSO) kinetic, and Elovich. The biosorption mechanism is predicted by fitting experimental data to the intraparticle diffusion model. The mathematical expressions of the kinetic models are listed in Equations (7.7) till (7.10):

PFO:

$$q_t = q_e\left(1 - e^{-k_1 t}\right) \tag{7.7}$$

PSO:

$$q_t = \frac{tk_2 q_e^{\,2}}{1 + tk_2 q_e} \tag{7.8}$$

Elovich:

$$q_t = \frac{1}{\beta}\ln(\alpha\beta) + \frac{1}{\beta}\ln(t) \tag{7.9}$$

Intraparticle diffusion:

$$q_t = K_p t^{0.5} + C \tag{7.10}$$

where k_1 (1/min) is the PFO rate constant, k_2 (g/mg min) is the PSO rate constant, β (mg/g) is the Elovich constant, α (mg/g min) is the initial adsorption rate, K_p (mg/g min$^{0.5}$) is the intraparticle diffusion rate constant, and C (mg/g) is the y-intercept of the intraparticle diffusion plot.

The biosorption process is assumed to be physisorption and the uptake rate is proportional to the availability of free binding sites if the adsorption kinetic obeys the PFO kinetic model (Lagergren, 1898). However, this model is appropriate to explain at the initial adsorption period of the biosorption, suggesting the mechanism is not limited by a concentration driving force, but by a physical interaction such as diffusion (Tran et al., 2017). On the other hand, PSO model describes the adsorption process to be chemisorption which is governed by mechanisms such as electron exchange or valency sharing between the adsorbent and adsorbate (Ho and McKay, 1998). The Elovich model is another kinetic model that is widely used to describe chemisorption system on a highly heterogeneous solid surface (Kumar et al., 2011).

In general, the kinetic models (PFO, PSO and Elovich) do not reveal more detailed biosorption mechanisms. Therefore, the intraparticle diffusion model is useful to identify the mechanism and predict the adsorption rate-limiting steps. Typically in the solid-liquid adsorption process, the biosorption rate is limited by external diffusion, surface diffusion, pore diffusion, or a combination of surface and pore diffusion (Tran et al., 2017). In short, the biosorption process is controlled by intraparticle diffusion if the q_t versus $t^{0.5}$ plot exhibits a linear relationship that passes through the origin. In contrast, multi-linearity in the plot suggests that the biosorption process is controlled by various mechanisms such as electrostatic interactions, interactions between surface functional groups and ionic exchange (Weber and Morris, 1963).

7.6 REVIEW ON BIOSORPTION PERFORMANCE AND MECHANISMS

Since biological material is complex, various mechanisms can be associated with a biosorption process. Biomasses consist of a variety of structural components or binding groups that act as the surface functional groups active in the biosorption process. Volesky (2007) reviewed the major binding groups and corresponding biomolecules for biosorption which are summarized in Table 7.2. In order to evaluate the performance of biosorption, knowledge of biochemistry and microbiology, in addition to thermodynamics and kinetics, is required.

In addition to functional groups, many other mechanisms govern the biosorption process. Table 7.3 summarizes the results of different biosorption studies on different pollutants conducted by researchers. The biosorption uptake of pollutants varied depending on the type of biomass used and the process parameters. In general, the biosorption mechanisms involved in pollutant decontamination are electrostatic interaction, chemisorption and hydrogen bonding.

Lam et al. (2016) reported that the biosorption of nickel (Ni^{2+}) on *Lansium domesticum* peel (LDP) biosorbent demonstrated multiple biosorption mechanisms. Electrostatic interaction mechanism was proposed as one of the responsible mechanisms. The pH_{PZC} for the LDP biosorbent was determined to be pH 5, implying that the biosorbent surface was positively charged at pH < 5, and it was negatively charged when pH > 5. At pH < 5, the Ni^{2+} removal was relatively low because of the strong electrostatic repulsion between the positively charged biosorbent surface and Ni^{2+}, whereas at pH > 5, the biosorbent surface deprotonated, causing the surface to be negatively charged which facilitated the electrostatic interaction in the biosorption uptake of Ni^{2+}. Notably, the pH control test was conducted without the presence of LDP biosorbent. By comparing the nickel speciation and pH control plot, it was revealed that the

TABLE 7.2
Major Binding Groups in Biosorption Process

Binding Groups	Component Structure	pK_a	Ligand Atom	Occurrence in Selected Biomolecules
Hydroxyl	–OH	9.5–13	O	PS, UA, SPS, AA
Carbonyl (ketone)	>C=O	—	O	Peptide bond
Carboxyl	–C=O │ OH	1.7–4.7	O	UA, AA
Sulfhydryl (thiol)	–SH	8.3–10.8	S	AA
Sulfonate	O ‖ –S=O ‖ O	1.3	O	SPS
Thioether	>S	—	S	AA
Amine	–NH_2	8–11	N	Cto, AA
Secondary amine	>NH	13	N	Cto, PG, peptide bond
Amide	–C=O │ NH_2	—	N	AA
Imine	=NH	11.6–12.6	N	AA
Imidazole	–C–N–H ‖ >CH H–C–N	6.0	N	AA
Phosphonate	OH │ –P=O │ OH	0.9–2.1, 6.1–6.8	O	PL
Phosphodiester	>P=O │ OH	1.5	O	TA, LPS

Source: Volesky, B., *Water Res.*, 41, 4017–4029, 2007.
PS = polysaccharides; UA = uronic acids; SPS = sulfated PS; Cto = chitosan; PG = peptidoglycan; AA = amino acids; TA = teichoic acid; PL = phospholipids; LPS = lipoPS.

predominant nickel species was Ni^{2+} below pH 8, while nickel hydroxide ($Ni(OH)_{2(s)}$) was the predominant species above pH 8. This indicated that the solubility of metal complexes decreased sufficiently above pH 8, causing precipitation which may complicate the biosorption process (Vijayaraghavan and Yun, 2008).

The biosorption of crystal violet on treated ginger waste (TGW) was governed by the influence of pH and the functional groups on the biosorbent surface. Fourier transform infrared spectroscopy (FTIR) analysis revealed that the TGW consisted of various functional groups, namely, hydroxyl and amine groups (3000–3500 cm^{-1}), C–H group (2926 cm^{-1}), COO (1662 cm^{-1}), N–H (1515 cm^{-1}), C–O (1156, 1457 cm^{-1}),

TABLE 7.3
Biosorption Performance and Mechanisms on Different Pollutants Removal

Biosorbent	Pollutants	q_m (mg/g)	Equilibrium Model	Kinetic Model	Mechanism	References
Synthetic Dyes						
Euchema Spinosum (marine macro-alga)	Methylene blue	833.33	Langmuir	PSO	Chemisorption, electrostatic interaction	Mokhtar et al. (2017)
Aspergillus carbonarius M333	Congo red	99.01	Langmuir	PSO	Chemisorption, electrostatic interaction	Bouras et al. (2017)
Penicillium glabrum Ph1mycelia	Congo red	101.01	Langmuir	PSO	Chemisorption, electrostatic interaction	Bouras et al. (2017)
Chemically modified masau stone	Orange II	136.8	Langmuir	PSO	Electrostatic attraction, hydrogen bonding	Albadarin et al. (2017)
Bacillus catenulatus JB-022 strain	Basic blue 3	139.74	Langmuir	PSO	Electrostatic interaction	Kim et al. (2015)
Thamnidium elegans	Reactive red 198	234.24	Langmuir	PSO	Electrostatic interaction	Akar et al. (2013)
Mixed fish scales	Acid blue 113	157.3	Langmuir	PSO	Electrostatic interaction, hydrogen bonding, hydrophobic interaction	Ooi et al. (2017)
Treated ginger waste	Crystal violet	277.7	Langmuir	PSO	Electrostatic interaction, hydrogen bonding, hydrophobic interaction	Kumar and Ahmad (2011)
Carbonized pomegranate peel	Malachite green	31.45	Langmuir	PFO	Physisorption, electrostatic interaction	Gündüz and Bayrak (2017)
Modified *Aspergillus versicolor*	Reactive black 5	227.27	Langmuir	PSO	Electrostatic interaction, hydrogen bonding	Huang et al. (2016)
Magnetic functionalized *Bacillus Subtilis*	Methylene blue	—	Freundlich	PSO	Electrostatic interaction, hydrogen bonding	Tural et al. (2017)
Earthworm manuare biochar	Rhodamine B	21.60	Langmuir	PSO	Ion exchange, hydrogen bonding, electrostatic interaction, π-π interaction	Wang et al. (2017)
Cetylpyridinium chloride- modified *Penicillium* YW 01	Acid blue 25	118.48	Langmuir	PSO	Electrostatic interaction	Yang et al. (2011)
Magnetic peach gum bead	Methylene blue	231.5	Langmuir	PSO	Electrostatic interaction	Li et al. (2018)

(*Continued*)

TABLE 7.3 (*Continued*)
Biosorption Performance and Mechanisms on Different Pollutants Removal

Biosorbent	Pollutants	q_m (mg/g)	Equilibrium Model	Kinetic Model	Mechanism	References
Heavy Metals						
Lansium domesticum peel	Ni^{2+}	10.1	Langmuir	PSO	Chemisorption, electrostatic interaction, surface complexion	Lam et al. (2016)
Hydrochloric acid modified durian peel	Zn^{2+}	36.73	Langmuir	PSO	Electrostatic interaction	Ngabura et al. (2018)
Parachlorella sp. Microalgae	Cd^{2+}	90.72	Langmuir	PFO	—	Dirbaz and Roosta (2018)
Palm oil sludge biochar	Pb^{2+}	—	Freundlich	PFO, PSO	Ionic exchange, complexation and precipitation	Lee et al. (2017)
Grapefruit peel	Cd^{2+}	42.09	Freundlich	PSO	Ionic exchange	Torab-Mostaedi et al. (2013)
	Ni^{2+}	46.13	Freundlich	PSO	Ionic exchange	Torab-Mostaedi et al. (2013)
Maugeotia gebuflexa biomass	As^{3+}	57.48	Langmuir	PSO	Ionic exchange, electrostatic interaction	Sarı et al. (2011)
Rape straw powder (shell)	Cu^{2+}	45.17	Langmuir	PSO	Ionic exchange, interaction with functional groups	Liu et al. (2018)
Hydrochloric acid–treated tomato waste	Cu^{2+}	34.48	Langmuir	PSO	Chemisorption, electrostatic interaction, hydrogen bonding	Yargıç et al. (2015)
Sugarcane bagasse	Hg^{2+}	35.7	Langmuir	PSO	—	Khoramzadeh et al. (2013)
Eichhornia crassipes	U^{6+}	142.85	Langmuir	PSO	Coordination, ionic exchange	Yi et al. (2016)
Pistachio hull waste	Cr^{6+}	116.3	Langmuir	PSO	Complexion, electrostatic interaction, chemisorption	Moussavi and Barikbin (2010)
Rapeseed biomass	Pb^{2+}	22.7	Langmuir	PSO	Physical and chemical interaction	Morosanu et al. (2017)
Modified rice straw biochar	Cd^{2+}	93.2	Langmuir	PSO	Chemisorption, electrostatic interaction	Zhang et al. (2018)
$ZnCl_2$-treated corn husk biochar	Pb^{2+}	—	Freundlich	PSO	Physisorption, ionic exchange, coordination, precipitation	Rwiza et al. (2018)

(*Continued*)

TABLE 7.3 (*Continued*)
Biosorption Performance and Mechanisms on Different Pollutants Removal

Biosorbent	Pollutants	q_m (mg/g)	Equilibrium Model	Kinetic Model	Mechanism	References
Other Pollutants						
Chemically modified eucalyptus sawdust biochar	Metronidazoles	167.5	Freundlich	PSO	Chemisorption, physisorption	Wan et al. (2016)
	Dimetridazole	200	Freundlich	PSO	Chemisorption, physisorption	Wan et al. (2016)
Microalgae (*Chlorella* sp.) derived biochar	*p*-nitrophenol	204.8	Freundlich	PSO	π-H bonding, electrostatic interaction	Zheng et al. (2017)
Crystalline nanocellulose from *Ulva lactuca*	Tetracycline	7.729	Redlich-Peterson	PSO	Cation exchange, hydrophobic interaction	Rathod et al. (2015)
Chemically modified *N*-biochar	Ibuprofen	90.46	Langmuir	PSO	Electrostatic interaction, hydrogen bonding	Mondal et al. (2016)
Chemically modified green alga (*Scenedesmus obliquus*)	Tramadol	140.25	Freundlich	PSO	Chemisorption	Ali et al. (2018)
Chitosan/alumina biocomposite	Nitrate	45.38	Freundlich	PSO	Electrostatic interaction	Golie and Upadhyayula (2017)
Alkali-steam treated elephant grass	Fluoride	7	—	—	C–F bond formation, hydrogen bonding	Manna et al. (2015)

q_m = the reported maximum biosorption capacity, PFO = pseudo-first-order, PSO = pseudo-second-order.

and C–N or sulphonic group (1047 cm^{-1}), which contributed to the biosorption process (Kumar and Ahmad, 2011). It was reported that chemical interaction between crystal violet molecules and carboxylate groups on the TGW surface was evident by the shift in FTIR spectra at carboxyl bands (1662 > 1649 cm^{-1} and 1457 > 1431 cm^{-1}). In addition, amine and hydroxyl groups of TGW were reported to be responsible for biosorbing the crystal violet dye molecules, as shown by the wavenumber change in N–H and O–H stretch region. This suggested the interaction between the nitrogen atom from amine groups of crystal violet and hydroxyl groups on the TGW surface which was supported by other research findings (Kannan et al., 2009; Jain and Jayaram, 2010). Notably, Kumar and Ahmad also suggested that electrostatic interaction of the biosorption process is as follows:

Electrostatic repulsion in acidic condition:

$$\text{TGW} + \text{H}^+ \rightarrow (\text{TGW})\ \text{H}^+$$

$$(\text{TGW})\ \text{H}^+ + \text{CV}^+ \rightarrow (\text{TGW})\ \text{H}^+ \leftrightarrow \text{CV}^+$$

Electrostatic attraction in basic condition:

$$\text{TGW} + \text{OH}^- \rightarrow (\text{TGW})\ \text{OH}^-$$

$$(\text{TGW})\ \text{OH}^- + \text{CV}^+ \rightarrow (\text{TGW})\ \text{OH}^- \bullet\bullet\bullet\bullet\bullet\ \text{CV}^+$$

They also proposed the possible mechanisms of the biosorption of crystal violet on TGW on the basis of FTIR spectra and active site analysis, as shown in Figure 7.3. The mechanisms that are responsible for crystal violet biosorption were van der Waals forces, hydrogen bonding between amine groups of crystal violet and TGW functional

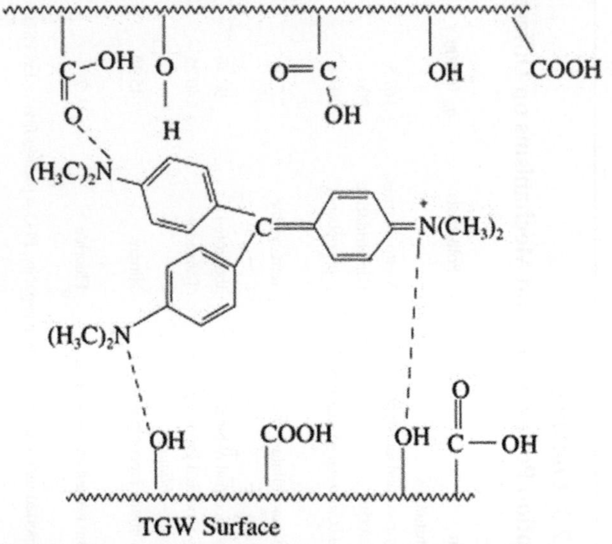

FIGURE 7.3 Proposed biosorption mechanisms on crystal violet removal by treated TGW. (From Kumar, R. and Ahmad, R., *Desalination*, 265, 112–118, 2011.)

groups, hydrophobic interaction at the hydrophobic parts of crystal violet and TGW, as well as electrostatic attraction at basic medium (Kumar and Ahmad, 2011).

Modification on the biosorbent not only alters its physicochemical properties but also directly affects the biosorption performance. Rwiza et al. (2018) have investigated the lead (Pb^{2+}) biosorption by physically and thermochemically modified corn and rice husk waste, that is, corn husk biochar (CHB), rice husk biochar (RHB), $ZnCl_2$-treated CHB (Zn-CHB), $ZnCl_2$-treated RHB, KOH-treated CHB (KOH-CHB) and KOH-treated RHB (KOH-RHB). The study showed that both raw corn husk (CH) and rice husk (RH) consisted of more functional groups than their modified derivatives. The presence of hydroxyl, carboxyl, phenol, alcohol and polysaccharide was found on all biosorbents except CHB. The abundance of functional groups (– CO, –COOH and –OH) could facilitate Pb^{2+} surface complexion (Lu et al., 2012). In chemical modification of CHB and RHB, the KOH and $ZnCl_2$ treatment enriched the cation exchange capacity for both CHB and RHB. This indicated that the chemical modification increased the availability of binding sites with oxygen functional groups (–OH and –COOH) on the biosorbent surface, which facilitated the biosorption uptakes of Pb^{2+} (Gaskin et al., 2008; Rwiza et al., 2018). Furthermore, the pH_{PZC} of CHB, RHB, Zn-CHB and Zn-RHB were 3.04, 3.04, 4.32 and 4.60, respectively, which was less than the working solution pH (pH 5). Below pH_{PZC}, the biosorbent surfaces were protonated, causing electrostatic repulsion with Pb^{2+}, whereas the biosorbent surfaces became negatively charged and promoted Pb^{2+} uptake when $pH > pH_{PZC}$. Another highlight from the study was that the precipitation of $Pb(OH)_2$ could occur when the working solution $pH > 6.5$ (Naiya et al., 2009). Therefore, it is important to select the working range of pH to deliver an efficient biosorption system.

Similarly, the thermochemical modification on CH and RH demonstrated a significant change in the physical properties on the developed biosorbents. After pyrolysis, the specific surface area of CHB and RHB increased by 2.5 times and 5.5 times, respectively, than the raw counterparts ($CH = 0.33$ m^2/g and $RH = 0.47$ m^2/g). Similarly, chemical treatment with $ZnCl_2$ and KOH significantly increased the specific surface area of Zn-CHB, Zn-RHB, KOH-CHB and KOH-RHB to 84.75, 203.62, 108.58, and 165.17 m^2/g, respectively, which were 250–430 greater than the raw biomasses. Scanning electron microscope (SEM) analysis revealed that CH and RH had fibrous and rough surfaces with irregular micropores, while modified biosorbents had clear honeycomb with microtubular (approximately 10 μm diameter) structures. The micropores in the modified biosorbents could contribute to the involvement of Pb^{2+} transport into the pores and fissures on the biosorbent surface as part of the biosorption mechanisms (Rwiza et al., 2018).

7.7 FUTURE PROSPECTS

Biosorption shows promising decontamination efficiency with relatively simple operation, low cost and environmental safety, indicating a potential efficient drinking water treatment technology for the future. Many processes have been patented for commercial application of biosorption. However, limited patents have been commercially established in wastewater treatment application. Fomina and Gadd (2014) summarized the list of patents related to biosorption process; their list is given in Table 7.4.

TABLE 7.4
List of Patents Related to Biosorption

Year	Title of the Invention	Patent Number
1973	Apparatus for the biological treatment of wastewater by the biosorption process	GB1324358
1973	Sorbent and method of manufacturing same	US3725291
1977	Process of treating mycelia of fungi for retention of metals	US4021368
1978	Method of treating a biomass	US4067821
1981	Microbiological recovery of metals	US4293333
1981	Process for recovering precious metals	US4289531
1982	Separation of uranium by biosorption	US4320093
1987	Process for the separation of metals from aqueous media	US4701261
1987	Treatment of microorganisms with alkaline solution to enhance metal uptake properties	US4690894
1987	Process for the separation of metals from aqueous media	US4701261
1988	Removal of contaminants	US4732681
1988	Biosorbent for gold	US4769223
1989	A process for the removal of thorium from raffinate	GB2228612A
1990	Metal recovery	US4898827
1990	Recovery of heavy and precious metals from aqueous solutions	WO9007468
1991	Removal of metal ions with immobilized metal ion-binding microorganisms	US5055402
1992	Process and apparatus for removing heavy metals from aqueous media by means of a bioadsorber	EP0475542
1992	Processes to recover and reconcentrate gold from its ores	US5152969
1992	Bioadsorption composition and process for production thereof	US5084389
1994	Ionic binding of microbial biomass	WO9413782
1994	Polymer beads containing an immobilized extractant for sorbing metals from solution	US5279745
1995	Method for adsorbing and separating heavy metal elements by using a tannin adsorbent and method of regenerating the adsorbent	US5460791
1996	Process for the removal of species containing metallic ions from effluents	US5538645
1996	Bead for removing dissolved metal contaminants	US5578547
1997	Polyaminosaccharide phosphate biosorbent	GB2306493
1997	Method for production of adsorption material	US5648313
1998	Biosorption system	WO9826851
1998	Biosorbent for heavy metals prepared from biomass	US5789204
1998	Bacteria expressing metallothionein gene into the periplasmic space, and method of using such bacteria in environment cleanup	US5824512
1998	Biosorption agents for metal ions and method for the production thereof	WO9848933
1998	Adsorption of PCB's using biosorbents	US5750065

(Continued)

TABLE 7.4 (*Continued*)
List of Patents Related to Biosorption

Year	Title of the Invention	Patent Number
1999	Hydrophilic urethane binder immobilizing organisms having active sites for binding noxious materials	US5976847
2000	Precipitating metals or degrading xenobiotic organic compounds with membrane immobilized microorganisms	US6013511
2000	Method for removing a heavy metal from sludge	US6027543
2001	Process for producing chitosan–glucan complexes, compounds producible therefrom and their use	US6333399
2002	Bioadsorption process for the removal of color from textile effluent	WO0242228
2002	Biosorption system	US6395143
2002	Adsorption means for radionuclides	US6402953
2003	Biosorbents and process for producing the same	US6579977
2003	Biocomposite (Biocer) for biosorption of heavy metals comprises an inorganic gel containing immobilized dry-stable cellular products	DE10146375A1
2004	Composite biosorbent for treatment of waste aqueous system(s) containing heavy metals	US6786336
2004	Heavy metal adsorbent composition	WO04022728
2006	A novel process for decolorization of colored effluents	WO06059348
2006	Process and plant for the removal of metals by biosorption from mining or industrial effluents	US20060070949
2007	Biosorption agents for metal ions and method for the production thereof	CA2282432C
2007	Petroleum biosorbent based on strains of bacteria and yeast	US20070202588
2008	Process for the removal of metals by biosorption from mining or industrial effluents	US7326344
2008	Biosorption system produced from biofilms supported in faujasite (FAU) Zeolite, process obtaining it and its usage for removal of hexavalent chromium (Cr(VI))	US20080169238
2010	Use of *Rhizopus stolonifer* (Ehrenberg) Vuillemin in methods for treating industrial wastewaters containing dyes	US7658849
2010	Use of *Cunninghamella elegans* Lendner in methods for treating industrial wastewaters containing dyes	US7790031
2011	Use of *Rhizomucor pusillus* (Lindt) Schipper in methods for treating industrial wastewaters containing dyes	US7935257
2011	*Pseudomonas alcaliphila* MBR and its application in bioreduction and biosorption	US0110269169
2011	Bacterial strain for a metal biosorption process	US7951578
2014	Biosorbents for the extraction of metals[a]	US8748153B2
2016	Adsorbents for biological removal of heavy metals[a]	CN104661964B/ WO2014012134A1

Source: Fomina, M. and Gadd, G.M., *Bioresource Technol.*, 160, 3–14, 2014.

[a] *The recent patent on biosorption.*

In addition, authors have listed two more recent patents in Table 7.4. Despite the increasing number of publications on the topic of biosorption, there is still a lack of insight about assessing the biosorption performance of various biosorbents on larger scale as most experiments are still conducted on the laboratory scale.

Innovative research in effective and practical biosorbent development is critical to fully utilize the great potential of biosorption processes on a commercial scale. Overall, the laboratory investigations highlighted the factors influencing the biosorption performance such as biosorption mechanisms, biosorbent selection, modification steps, process parameters and mode of operation. Investigation of the biosorption efficiency on real industrial effluent and operating in continuous mode is necessary to accurately simulate and evaluate the biosorption performance in real-life applications.

Understanding the synthesis and functionalization mechanism in biosorbent development is important in developing an effective and feasible biosorbent that is versatile and highly resistance against extreme process conditions for future application. With the current understanding and future progression, it is expected that biosorption will offer beneficial prospects in future wastewater treatment technology to provide significant contributions to water pollution control.

7.8 CONCLUSION

Investigations have demonstrated the promising potential of biosorption processes as future wastewater treatment technology. The performance of biosorption process is controlled by various factors such as selection of biosorbent, modification and functionalization steps, process parameters and biosorption mechanisms. The biosorption system can be described by several empirical models (equilibrium isotherm and kinetic models) to design and optimize system performance. However, there is still a lack of in-depth understanding related to biosorbent development, and biosorption performance in continuous mode and multicomponent system, which hinders the full commercialization of this technology. Therefore, more innovative investigations should be conducted to provide insights on biosorption, leading to the development of next-generation biosorption treatment systems.

REFERENCES

Akar, T., S. Arslan, and S. T. Akar. 2013. "Utilization of thamnidium elegans fungal culture in environmental cleanup: A reactive dye biosorption study." *Ecological Engineering* 58: 363–370. doi:10.1016/j.ecoleng.2013.06.026.

Albadarin, A. B., S. Solomon, T. A. Kurniawan, C. Mangwandi, and G. Walker. 2017. "Single, simultaneous and consecutive biosorption of Cr(VI) and Orange II onto chemically modified masau stones." *Journal of Environmental Management* 204: 365–374. doi:10.1016/j.jenvman.2017.08.042.

Ali, M. E. M., A. M. El-Aty, M. I. Badawy, and R. K. Ali. 2018. "Removal of pharmaceutical pollutants from synthetic wastewater using chemically modified biomass of green alga Scenedesmus obliquus." *Ecotoxicology and Environmental Safety* 151: 144–152. doi:10.1016/j.ecoenv.2018.01.012.

Bouras, H. D., A. R. Yeddou, N. Bouras, D. Hellel, M. D. Holtz, N. Sabaou, A. Chergui, and B. Nadjemi. 2017. "Biosorption of Congo red dye by Aspergillus carbonarius M333 and Penicillium glabrum Pg1: Kinetics, equilibrium and thermodynamic studies." *Journal of the Taiwan Institute of Chemical Engineers* 80: 915–923. doi:10.1016/j.jtice.2017.08.002.

Daneshfozoun, S., M. A. Abdullah, and B. Abdullah. 2017. "Preparation and characterization of magnetic biosorbent based on oil palm empty fruit bunch fibers, cellulose and Ceiba pentandra for heavy metal ions removal." *Industrial Crops and Products* 105: 93–103. doi:10.1016/j.indcrop.2017.05.011.

Dirbaz, M., and A. Roosta. 2018. "Adsorption, kinetic and thermodynamic studies for the biosorption of cadmium onto microalgae Parachlorella sp." *Journal of Environmental Chemical Engineering* 6(2): 2302–2309. doi:10.1016/j.jece.2018.03.039.

Dubinin, M. M., and L. V. Radushkevich. 1947. Equation of the Characteristic Curve of Activated Charcoal. *Proceedings of the Academy of Sciences of the USSR, Physical Chemistry Section* 55: 331–333.

Fomina, M., and G. M. Gadd. 2014. "Biosorption: Current perspectives on concept, definition and application." *Bioresource Technology* 160: 3–14. doi:10.1016/j.biortech.2013.12.102.

Freundlich, H. 1906. "Über die absorption in lösungen." *Über Die Adsorption in Lösungen* 385–470.

Gadd, G. M. 2009. "Biosorption: Critical review of scientific rationale, environmental importance and significance for pollution treatment." *Journal of Chemical Technology and Biotechnology* 84(1): 13–28. doi:10.1002/jctb.1999.

Gaskin, J. W., C. Steiner, K. Harris, K. C. Das, and B. Bibens. 2008. "Effect of low-temperature pyrolysis conditions on biochar for agricultural use." *Transactions of the ASABE* 51(6): 2061–2069.

Golie, W. M., and S. Upadhyayula. 2017. "An investigation on biosorption of nitrate from water by chitosan based organic-inorganic hybrid biocomposites." *International Journal of Biological Macromolecules* 97: 489–502. doi:10.1016/j.ijbiomac.2017.01.066.

Gündüz, F., and B. Bayrak. 2017. "Biosorption of malachite green from an aqueous solution using pomegranate peel: Equilibrium modelling, kinetic and thermodynamic studies." *Journal of Molecular Liquids* 243: 790–798. doi:10.1016/j.molliq.2017.08.095.

Gupta, N. K., A. Sengupta, A. Gupta, J. R. Sonawane, and H. Sahoo. 2018. "Biosorption-an alternative method for nuclear waste management: A critical review." *Journal of Environmental Chemical Engineering* 6(2): 2159–2175. doi:10.1016/j.jece.2018.03.021.

Hall, K. R., L. C. Eagleton, A. Acrivos, and T. Vermeulen. 1966. "Pore- and solid-diffusion kinetics in fixed-bed adsorption under constant-pattern conditions." *Industrial and Engineering Chemistry Fundamentals* 5(2): 212–223. doi:10.1021/i160018a011.

Hiew, B. Y. Z., L. Y. Lee, X. J. Lee, S. Thangalazhy-Gopakumar, S. Gan, S. S. Lim, G.-T. Pan, T. C.K. Yang, W. S. Chiu, and P. S. Khiew. 2018. "Review on synthesis of 3D graphene-based configurations and their adsorption performance for hazardous water pollutants." *Process Safety and Environmental Protection* 116: 262–286. doi:10.1016/j.psep.2018.02.010.

Ho, Y. S., and G. McKay. 1998. "Sorption of dye from aqueous solution by peat." *Chemical Engineering Journal* 70(2): 115–124. doi:10.1016/S1385-8947(98)00076-X.

Huang, J., D. Liu, J. Lu, H. Wang, X. Wei, and J. Liu. 2016. "Biosorption of reactive black 5 by modified Aspergillus versicolor biomass: Kinetics, capacity and mechanism studies." *Colloids and Surfaces A: Physicochemical and Engineering Aspects* 492: 242–248. doi:10.1016/j.colsurfa.2015.11.071.

Jain, S., and R. V. Jayaram. 2010. "Removal of basic dyes from aqueous solution by low-cost adsorbent: Wood apple shell (Feronia acidissima)." *Desalination* 250(3): 921–927. doi:10.1016/j.desal.2009.04.005.

Kannan, C., N. Buvaneswari, and T. Palvannan. 2009. "Removal of plant poisoning dyes by adsorption on Tomato Plant Root and green carbon from aqueous solution and its recovery." *Desalination* 249(3): 1132–1138. doi:10.1016/j.desal.2009.06.042.

Khoramzadeh, E., B. Nasernejad, and R. Halladj. 2013. "Mercury biosorption from aqueous solutions by Sugarcane Bagasse." *Journal of the Taiwan Institute of Chemical Engineers* 44(2): 266–269. doi:10.1016/j.jtice.2012.09.004.

Kim, S. Y., M. R. Jin, C. H. Chung, Y.-S. Yun, K. Y. Jahng, and K.-Y. Yu. 2015. "Biosorption of cationic basic dye and cadmium by the novel biosorbent Bacillus catenulatus JB-022 strain." *Journal of Bioscience and Bioengineering* 119(4): 433–439. doi:10.1016/j.jbiosc.2014.09.022.

Kumar, P. S., S. Ramalingam, S. D. Kirupha, A. Murugesan, T. Vidhyadevi, and S. Sivanesan. 2011. "Adsorption behavior of nickel(II) onto cashew nut shell: Equilibrium, thermodynamics, kinetics, mechanism and process design." *Chemical Engineering Journal* 167(1): 122–131. doi:10.1016/j.cej.2010.12.010.

Kumar, R., and R. Ahmad. 2011. "Biosorption of hazardous crystal violet dye from aqueous solution onto treated ginger waste (TGW)." *Desalination* 265(1): 112–118. doi:10.1016/j.desal.2010.07.040.

Kumar, R., R. K. Sharma, and A. P. Singh. 2017. "Cellulose based grafted biosorbents—Journey from lignocellulose biomass to toxic metal ions sorption applications—A review." *Journal of Molecular Liquids* 232: 62–93. doi:10.1016/j.molliq.2017.02.050.

Lagergren, S. 1898. "Handlingar." *Band* 24(4): 1–39.

Lam, Y. F., L. Y. Lee, S. J. Chua, S. S. Lim, and S. Gan. 2016. "Insights into the equilibrium, kinetic and thermodynamics of nickel removal by environmental friendly Lansium domesticum peel biosorbent." *Ecotoxicology and Environmental Safety* 127: 61–70. doi:10.1016/j.ecoenv.2016.01.003.

Langmuir, I. 1918. "The adsorption of gases on plane surfaces of glass, mica and platinum." *Journal of the American Chemical Society* 40(9): 1361–1403. doi:10.1021/ja02242a004.

Lee, X. J., L. Y. Lee, B. Y. Hiew, S. Gan, S. Thangalazhy-Gopakumar, and H. K. Ng. 2017. "Multistage optimizations of slow pyrolysis synthesis of biochar from palm oil sludge for adsorption of lead." *Bioresource Technology* 245: 944–953. doi:10.1016/j.biortech.2017.08.175.

Li, C., X. Wang, D. Meng, and L. Zhou. 2018. "Facile synthesis of low-cost magnetic biosorbent from peach gum polysaccharide for selective and efficient removal of cationic dyes." *International Journal of Biological Macromolecules* 107: 1871–1878. doi:10.1016/j.ijbiomac.2017.10.058.

Liu, X., Z.-Q. Chen, B. Han, C.-L. Su, Q. Han, and W.-Z. Chen. 2018. "Biosorption of copper ions from aqueous solution using rape straw powders: Optimization, equilibrium and kinetic studies." *Ecotoxicology and Environmental Safety* 150: 251–259. doi:10.1016/j.ecoenv.2017.12.042.

Long, J., X. Gao, M. Su, H. Li, D. Chen, and S. Zhou. 2018. "Performance and mechanism of biosorption of nickel(II) from aqueous solution by non-living Streptomyces roseorubens SY." *Colloids and Surfaces A: Physicochemical and Engineering Aspects* 548: 125–133. doi:10.1016/j.colsurfa.2018.03.040.

Lu, H., W. Zhang, Y. Yang, X. Huang, S. Wang, and R. Qiu. 2012. "Relative distribution of Pb 2+ sorption mechanisms by sludge-derived biochar." *Water Research* 46(3): 854–862. doi:10.1016/j.watres.2011.11.058.

Manna, S., D. Roy, P. Saha, and B. Adhikari. 2015. "Defluoridation of aqueous solution using alkali–steam treated water hyacinth and elephant grass." *Journal of the Taiwan Institute of Chemical Engineers* 50: 215–222. doi:10.1016/j.jtice.2014.12.003.

Mokhtar, N., E. A. Aziz, A. Aris, W. F. W. Ishak, and N. S. M. Ali. 2017. "Biosorption of azo-dye using marine macro-alga of Euchema Spinosum." *Journal of Environmental Chemical Engineering* 5(6): 5721–5731. doi:10.1016/j.jece.2017.10.043.

Mondal, S., K. Aikat, and G. Halder. 2016. "Biosorptive uptake of ibuprofen by chemically modified Parthenium hysterophorus derived biochar: Equilibrium, kinetics, thermodynamics and modeling." *Ecological Engineering* 92: 158–172. doi:10.1016/j.ecoleng.2016.03.022.

Morosanu, I., C. Teodosiu, C. Paduraru, D. Ibanescu, and L. Tofan. 2017. "Biosorption of lead ions from aqueous effluents by rapeseed biomass." *New Biotechnology* 39: 110–124. doi:10.1016/j.nbt.2016.08.002.

Moscatello, N., G. Swayambhu, C. H. Jones, J. Xu, N. Dai, and B. A. Pfeifer. 2018. "Continuous removal of copper, magnesium, and nickel from industrial wastewater utilizing the natural product yersiniabactin immobilized within a packed-bed column." *Chemical Engineering Journal* 343: 173–179. doi:10.1016/j.cej.2018.02.093.

Moussavi, G., and B. Barikbin. 2010. "Biosorption of chromium(VI) from industrial wastewater onto pistachio hull waste biomass." *Chemical Engineering Journal* 162(3): 893–900. doi:10.1016/j.cej.2010.06.032.

Naiya, T. K., A. K. Bhattacharya, S. Mandal, and S. K. Das. 2009. "The sorption of lead(II) ions on rice husk ash." *Journal of Hazardous Materials* 163(2–3): 1254–1264. doi:10.1016/j.jhazmat.2008.07.119.

Ngabura, M., S. A. Hussain, W. A. Ghani, M. S. Jami, and Y. P. Tan. 2018. "Utilization of renewable durian peels for biosorption of zinc from wastewater." *Journal of Environmental Chemical Engineering* 6(2): 2528–2539. doi:10.1016/j.jece.2018.03.052.

Ooi, J., L. Y. Lee, B. Y. Z. Hiew, S. Thangalazhy-Gopakumar, S. S. Lim, and S. Gan. 2017. "Assessment of fish scales waste as a low cost and eco-friendly adsorbent for removal of an azo dye: Equilibrium, kinetic and thermodynamic studies." *Bioresource Technology* 245: 656–664. doi:10.1016/j.biortech.2017.08.153.

Rathod, M., S. Haldar, and S. Basha. 2015. "Nanocrystalline cellulose for removal of tetracycline hydrochloride from water via biosorption: Equilibrium, kinetic and thermodynamic studies." *Ecological Engineering* 84: 240–249. doi:10.1016/j.ecoleng.2015.09.031.

Rwiza, M. J., S.-Y. Oh, K.-W. Kim, and S. D. Kim. 2018. "Comparative sorption isotherms and removal studies for Pb(II) by physical and thermochemical modification of low-cost agro-wastes from Tanzania." *Chemosphere* 195: 135–145. doi:10.1016/j.chemosphere.2017.12.043.

Saha, G. C., M. I. U. Hoque, M. A. M. Miah, R. Holze, D. A. Chowdhury, S. Khandaker, and S. Chowdhury. 2017. "Biosorptive removal of lead from aqueous solutions onto Taro (*Colocasiaesculenta* (L.) Schott) as a low cost bioadsorbent: Characterization, equilibria, kinetics and biosorption-mechanism studies." *Journal of Environmental Chemical Engineering* 5(3): 2151–2162. doi:10.1016/j.jece.2017.04.013.

Sarı, A., Ö. D. Uluozlü, and M. Tüzen. 2011. "Equilibrium, thermodynamic and kinetic investigations on biosorption of arsenic from aqueous solution by algae (*Maugeotia genuflexa*) biomass." *Chemical Engineering Journal* 167(1): 155–161. doi:10.1016/j.cej.2010.12.014.

Temkin, M. and V. Phyzev. 1940. "Kinetics of ammonia synthesis on promoted iron catalysts." *Acta Physicochimica* 12: 327–356.

Tka, N., M. Jabli, T. A. Saleh, and G. A. Salman. 2018. "Amines modified fibers obtained from natural Populus tremula and their rapid biosorption of Acid Blue 25." *Journal of Molecular Liquids* 250: 423–432. doi:10.1016/j.molliq.2017.12.026.

Torab-Mostaedi, M., M. Asadollahzadeh, A. Hemmati, and A. Khosravi. 2013. "Equilibrium, kinetic, and thermodynamic studies for biosorption of cadmium and nickel on grapefruit peel." *Journal of the Taiwan Institute of Chemical Engineers* 44(2): 295–302. doi:10.1016/j.jtice.2012.11.001.

Tran, H. N., S.-J. You, A. Hosseini-Bandegharaei, and H.-P. Chao. 2017. "Mistakes and inconsistencies regarding adsorption of contaminants from aqueous solutions: A critical review." *Water Research* 120: 88–116. doi:10.1016/j.watres.2017.04.014.

Tural, B., E. Ertaş, B. Enez, S. A. Fincan, and S. Tural. 2017. "Preparation and characterization of a novel magnetic biosorbent functionalized with biomass of Bacillus Subtilis: Kinetic and isotherm studies of biosorption processes in the removal of Methylene Blue." *Journal of Environmental Chemical Engineering* 5(5): 4795–4802. doi:10.1016/j.jece.2017.09.019.

Ungureanu, G., S. C. R. Santos, I. Volf, R. A. R. Boaventura, and C. M. S. Botelho. 2017. "Biosorption of antimony oxyanions by brown seaweeds: Batch and column studies." *Journal of Environmental Chemical Engineering* 5(4): 3463–3471. doi:10.1016/j.jece.2017.07.005.

Vijayaraghavan, K., and R. Balasubramanian. 2015. "Is biosorption suitable for decontamination of metal-bearing wastewaters? A critical review on the state-of-the-art of biosorption processes and future directions." *Journal of Environmental Management* 160: 283–296. doi:10.1016/j.jenvman.2015.06.030.

Vijayaraghavan, K., and Y.-S. Yun. 2008. "Bacterial biosorbents and biosorption." *Biotechnology Advances* 26(3): 266–291. doi:10.1016/j.biotechadv.2008.02.002.

Volesky, B. 2001. "Detoxification of metal-bearing effluents: Biosorption for the next century." *Hydrometallurgy* 59(2–3): 203–216. doi:10.1016/S0304-386X(00)00160-2.

Volesky, B. 2007. "Biosorption and me." *Water Research* 41(18): 4017–4029. doi:10.1016/j.watres.2007.05.062.

Wan, S., Z. Hua, L. Sun, X. Bai, and L. Liang. 2016. "Biosorption of nitroimidazole antibiotics onto chemically modified porous biochar prepared by experimental design: Kinetics, thermodynamics, and equilibrium analysis." *Process Safety and Environmental Protection* 104: 422–435. doi:10.1016/j.psep.2016.10.001.

Wang, Z., D. Shen, F. Shen, C. Wu, and S. Gu. 2017. "Kinetics, equilibrium and thermodynamics studies on biosorption of Rhodamine B from aqueous solution by earthworm manure derived biochar." *International Biodeterioration & Biodegradation* 120: 104–114. doi:10.1016/j.ibiod.2017.01.026.

Weber, W. J., and J. C. Morris. 1963. "Kinetics of adsorption of carbon from solution." *American Society of Civil Engineering* 89: 31–60.

Wen, X., C. Du, G. Zeng, D. Huang, J. Zhang, L. Yin, S. Tan et al. 2018. "A novel biosorbent prepared by immobilized Bacillus licheniformis for lead removal from wastewater." *Chemosphere* 200: 173–179. doi:10.1016/j.chemosphere.2018.02.078.

Xu, H., L. Tan, H. Cui, M. Xu, Y. Xiao, H. Wu, H. Dong, X. Liu, G. Qiu, and J. Xie. 2018. "Characterization of Pd(II) biosorption in aqueous solution by Shewanella oneidensis MR-1." *Journal of Molecular Liquids* 255: 333–340. doi:10.1016/j.molliq.2018.01.168.

Yagub, M. T., T. K. Sen, S. Afroze, and H. M. Ang. 2014. "Dye and its removal from aqueous solution by adsorption: A review." *Advances in Colloid and Interface Science* 209: 172–184. doi:10.1016/j.cis.2014.04.002.

Yang, Y., D. Jin, G. Wang, D. Liu, X. Jia, and Y. Zhao. 2011. "Biosorption of Acid Blue 25 by unmodified and CPC-modified biomass of Penicillium YW01: Kinetic study, equilibrium isotherm and FTIR analysis." *Colloids and Surfaces B: Biointerfaces* 88(1): 521–526. doi:10.1016/j.colsurfb.2011.07.047.

Yargıç, A. Ş, R. Z. Yarbay Şahin, N. Özbay, and E. Önal. 2015. "Assessment of toxic copper(II) biosorption from aqueous solution by chemically-treated tomato waste." *Journal of Cleaner Production* 88: 152–159. doi:10.1016/j.jclepro.2014.05.087.

Yi, Z.-J., J. Yao, H.-L. Chen, F. Wang, Z.-M. Yuan, and X. Liu. 2016. "Uranium biosorption from aqueous solution onto Eichhornia crassipes." *Journal of Environmental Radioactivity* 154: 43–51. doi:10.1016/j.jenvrad.2016.01.012.

Zhang, H., X. Yue, F. Li, R. Xiao, Y. Zhang, and D. Gu. 2018. "Preparation of rice straw-derived biochar for efficient cadmium removal by modification of oxygen-containing functional groups." *Science of the Total Environment* 631–632: 795–802. doi:10.1016/j.scitotenv.2018.03.071.

Zhang, Q., T. Lu, D.-M. Bai, D.-Q. Lin, and S.-J. Yao. 2016. "Self-immobilization of a magnetic biosorbent and magnetic induction heated dye adsorption processes." *Chemical Engineering Journal* 284: 972–978. doi:10.1016/j.cej.2015.09.047.

Zheng, H., W. Guo, S. Li, Y. Chen, Q. Wu, X. Feng, R. Yin, S.-H. Ho, N. Ren, and J.-S. Chang. 2017. "Adsorption of p-nitrophenols (PNP) on microalgal biochar: Analysis of high adsorption capacity and mechanism." *Bioresource Technology* 244: 1456–1464. doi:10.1016/j.biortech.2017.05.025.

8 Liquid-Liquid Separation

Hui Yi Leong, Pau Loke Show, K. Vogisha Kunjunee, Qi Wye Neoh, and Payal Sunil Thadani

8.1 INTRODUCTION

Liquid-liquid separation is a process used to isolate different constituents in a liquid mixture. A very simple form of separation involves bringing an immiscible or partially miscible solvent into contact with a mixture in a countercurrent manner. The solvent separates the liquid solution by removing the solute from it. Realistically, there tends to be a multitude of solutes within any given solution, and so a variety of solvents may need to be mixed together to make a single, highly selective solvent that is capable of removing several different solutes depending on their chemical structure.

Liquid-liquid separation is a well-established process that has been widely used industrially since it was introduced into the petrochemical industry in 1909. Lazar Edeleanu devised a process that enabled aromatic hydrocarbons to be extracted from kerosene with the use of sulphur dioxide as a solvent (Leffler, 2008). Over time, liquid-liquid separation has become a key operation in a wide range of industries, such as biochemical, metallurgical, pharmaceutical, food processing and nuclear industries globally, in addition to the petrochemical industry where it was first applied (Grandison and Lewis, 1996). The primary mechanism in this operation is the mass transfer of the solute from the mixture to the solvent, from one liquid phase into the other. The separation is optimal when there is a large interfacial area for the component to diffuse between the two liquids. Therefore, separators are designed to create dispersion of droplets of one liquid within the other continuous phase (Grandison and Lewis, 1996). On one hand, where distillation uses the wide range of volatilities of each component in a mixture, liquid-liquid separation utilizes the relative solubility that enables the components to split and disperse into a second component, namely, the immiscible solvent. Generally, this separation method is preferred when the constituents in a mixture are sensitive to heat or are nonvolatile. Distillation would be an inadequate operation for such cases because the components might react or degrade at high temperatures and form by-products, or they may form azeotropes (Qureshi et al., 2014). These factors make liquid-liquid separation the ideal operation for the biochemical engineering industry.

On the other hand, the liquid biphasic partitioning system (LBPS) is a liquid-liquid separation that uses a biphasic system for the separation and partition of biomolecules. LBPS, which is commonly known as the aqueous two-phase system (ATPS), was discovered by Martinus Willem Beijerinck in 1896, where he observed and reported the phenomenon of two distinct phase formations that left a mixed agar and gelatin settled by gravity (Iqbal et al., 2016). However, no further research was done to study the phenomenon he had reported, until close to a century later, in 1986. Albertsson attempted to purify chloroplast by applying ATPS. The attempt was successful, and Albertsson published his first paper on the application of ATPS in the extraction and purification of biomolecules (Grilo et al., 2014) that is still studied extensively today. This discovery was a major breakthrough in biomolecules separation because the conditions used in LBPS are often attributed to high water content, ambient operating conditions, and the low toxicity of phase-forming components, which are very suitable for extraction of biomolecules compared to conventional liquid-liquid separation, which uses a large amount of toxic organic solvents and harsh operating conditions (Iqbal et al., 2016), has a high tendency to denature proteins, and so on. Upon its discovery, LBPS consisted mainly of combinations between polymers and salts, such as polymer/polymer, polymer/salt, and salt/salt. As the extraction takes place in LBPS, two phases are formed. The top phase is enriched with solutes, and the bottom phase is left with the remains of fermented broths or parent samples (Freire et al., 2012). Later developments in LBPS include the implementation of ionic liquids, alcohols and thermos separating polymers.

Because biomolecules are susceptive in nature, selecting an ideal purification method is very complex. Liquid biphasic flotation system (LBFS), commonly known as aqueous two-phase flotation (ATPF), is an integration process of LBPS and the solvent sublation (SS) method. SS is employed in the recovery of surface-active and hydrophobic compounds, and has showed impressive efficiency in the separation of biomolecules. LBFS is highly biocompatible, has a high enrichment factor and low environment impact, and is inexpensive and simple to operate. The recovery of phase forming molecules is easier with this method. For example, in 2009, Bi and his colleagues proposed the ATPF method in the separation and concentration of penicillin G from fermentation broth in Polyethylene glycol (PEG)/ammonium sulphate biphasic system. The authors discovered a 97% rate of separation efficiency and 19 of concentration coefficient after the flotation process (Bi et al., 2009).

8.2 CONVENTIONAL LIQUID-LIQUID SEPARATION METHODS

8.2.1 Variables Affecting Separation

Within every liquid-liquid separation unit is a carrier stream that includes all the components that must be recovered when the solvent comes into contact with it. Six factors must be carefully considered when choosing a solvent for separation (Table 8.1).

8.2.2 Current Technologies

Many technologies are currently available for conventional liquid-liquid separation, and they can be divided into two categories: batch and continuous. Batch separation tends to be used for smaller, higher precision, valuable separations like pharmaceuticals, whereas continuous processes are more economical for processes with large throughputs, that is, chemicals and petrochemicals.

TABLE 8.1
Solvent Selection Factors

Factor	Description
Miscibility	It is key that these two liquids are immiscible because without this difference in miscibility, a dispersion would not form, and the rate of mass transfer would be very slow. For this reason, aqueous solvents are used for the separation of organic components, and vice versa.
Density	Another property that must be taken into consideration when choosing a solvent is its density and how that compares to the density of the mixture. Fluids with different densities separate more quickly and easily. Depending on the difference in densities, droplets form and collect either above or below the continuous stream, at the interface of the two phases.
Distribution coefficient (*K*)	Mass transfer is made easier by the varying compositions in both the phases. The distribution coefficient of the components that need to be separated is symbolized by *K*. This coefficient provides a mathematical value of the distribution of an organic compound near the interface, within the aqueous and organic liquids. It is simply a ratio of the solute dissolving in the organic phase to dissolving in the aqueous phase, as shown by the following equation: $K = X_\alpha/X_\beta$. Here *X* denotes the ratio of mole fractions in each phase, and organic and aqueous are symbolized by α and β, respectively.
Partition ratio	In a separating vessel, after the feed liquid and solvent have settled, they separate into layers. The continuous stream that contains the solute that needs to be removed is often called the raffinate, and the solution containing the separated compounds is called the extract. Hence, the partition ratio is simply the ratio of weight fraction of the solute in the extract and the raffinate. This is an important factor that aids in determining the number of stages required in the unit and the volume of solvent required to achieve a desired separation.
Safety	This factor is paramount in ensuring that the likelihood of incidents is minimized by selecting solvents that are nontoxic and inflammable.
Cost	As important as this factor is, the solvent should never be purchased without considering the other factors, too.

Source: Towler, G. and Sinnott, R. K., Random packing, in *Chemical Engineering Design—Principles, Practice and Economics of Plant and Process Design*, 2nd ed., Butterworth Heinemann, Amsterdam, the Netherlands, 2012.

8.2.2.1 Mixer-Settlers

A mixer-settler is a batch vessel that consists of two stages. The first chamber is an area where the two phases are mixed thoroughly, and the second chamber is where the mixture settles. The two phases separate by the force of gravity. In the first phase, the presence of a mechanical impeller brings the solvent and the feed stream into close contact with each other to permit the solutes to transfer from one phase to the other. The agitator's motor mechanism also drives a turbine that pumps the now-homogenized mixture into the settling tank. The size of mixer-settlers varies with their purpose; for instance, laboratory mixers contain just the primary mixing stage, but mixers used in the copper-processing industry require a minimum of three mixers, and each of these acts as both a mixer and a pump. The benefits of using mixer-settlers in series is that the mixing time is prolonged, ensuring that all the material mixes thoroughly and there is no opportunity for the unmixed material to end up in the settling tank.

Once the two phases are in the settling compartment, static decantation allows the two phases to separate with the aid of coalescence plates (second phase). The phases are split according to their density and flow over their respective weirs, which can be adjusted at different heights to allow liquids of different densities to flow over and be separated from the other liquid.

This technology can be found in many industries, ranging from the pharmaceutical to the nuclear industries. Mixer-settlers are used over other liquid-liquid separation technologies as they have an extremely high extraction yield and can process high flow rates. The equipment is simple and can be easily controlled, making it less susceptible to malfunctions. Mixer-settlers are ideal for mixtures that require a large settling time. An example of such a separation is the continuous protein purification using a mixer-settler containing four chambers, with perfluorocarbon as a solvent mixed with the protein solution (Cleveland and Morris, 2014).

8.2.2.2 Packed Extraction Towers

Separation towers can have many different arrangements within, for example, packing, rotating disks, and sieve trays, or they can even be empty but contain spray feeds. Generally, packed columns are the most efficient compared to the rest because of the larger contact area available and the reduced flow rate of the continuous phase within, which increases the rate of mass transfer. It also enables the droplets from either the heavier or lighter phase to coalesce and provides a more even distribution of droplets throughout the column. It is vital that the packing should be sufficiently coated with the down-flowing liquid to prevent back-mixing and hinder recirculation. The flows within the tower countercurrent limit axial dispersion within the tower.

The packing used is available in many shapes and sizes; the most common are the Raschig Ring and saddle-shaped configurations. The packing can be randomly dumped into the column, creating irregular flow patterns that decrease the velocity of the dispersed phase. Or the towers can contain structural packing, which are similar to that used in distillation, but they can create rivulets and areas of localized flow paths.

Packed columns, although less predominant than mixer-settlers, have more specific uses in industry because they can handle a high-specific throughput per unit of cross-sectional area in terms of both the phases. They are easy to maintain, and their design does not contain any moving parts, making them easier to operate. The towers can also be constructed many different materials, so they are durable and withstand corrosion, especially if handling corrosive material, at high temperatures and pressures.

One of the most common uses of packed separation towers is the neutralization of waste acids. Another common use of packed towers is the decaffeination of coffee liquor because they do not generate foam, which is a problem when a mixer-settler is used for the extraction of caffeine. In a packed column, the down-flowing coffee liquor forms thin films on the random packing, and it is brought into contact with the supercritical carbon dioxide that is flowing in the direction opposite to it. Mass transfer occurs, and the caffeine is separated from the liquor and into the carbon dioxide stream, which exits the column at the top. It is then washed using water as a solvent in a second packed separation column. The water is evaporated from the caffeine solution, giving a pure and dry caffeine product (Couper et al., 2012; Mchugh and Krukonis, 2013).

8.2.2.3 Centrifugal Separators

Centrifugal separators contain a rotor inside the casing of a centrifuge, and the movement of its rotation causes the mixing of two liquids that are otherwise immiscible. The two liquids are separated due to their differences in density and by the field of gravity within the vessel. This continuous separation process allows for the removal of a component from the feed into the solvent. The two streams enter from the top of the vessel through different inlets and into an annular area within the rotor. Here the two liquids are mixed, depending on the speed of rotation, the size and distribution of the droplets and the interfacial area available; the mass transfer occurs between the two liquids. Within the casing are baffles that prevent a vortex of fluid from forming in the rotor, and they direct the fluid down into the exit through the bottom of the vessel.

The force generated by the rotation within the rotor causes the phases to separate very quickly as they tend to experience strong centrifugal forces that cause them to flow upward in the rotor and split into layers because of their respective densities. They are then split off and are removed as they flow over a specific weir into a collection tank. This machine performs mixing, centrifugation, pumping, and separation with the use of only one rotating part, which makes it a very compact and energy-efficient device.

It is primarily used in separating oil from water as both substances have different densities. The oil droplets that are lighter tend to gather around the center of the rotor where the weir for oil removal is, whereas the water is collected in the outer radius. This technology is capable of producing water that has a concentration range of 50–70 ppm of oil (de Haan and Bosch, 2013; Harker et al., 2013).

8.2.2.4 Other Separation Towers

1. *Perforated plate towers*: Much like the packed columns, perforated plate towers aim to reduce the flow rate of the down-coming liquid with the intention of increasing mass transfer between the dispersed droplets that form on the plates and the solvent. The dispersed phase is spread out over the height of the tower as it traverses through the perforated plates and tends to gather on top of or underneath the plates. Eventually, after reaching a large enough hydrostatic pressure, the liquid is able to pass through the holes, causing it to split up into tiny droplets that merge once again after reaching the plate below. Generally, the perforations in the plate are between 2 and 10 mm and the distance between the plates can be within the range of 0.1–0.5 m. This means that most columns have approximately 20 plates and have an average

diameter of 4 m. The advantages of perforated plate towers are their flexible operation, their low repair rates due to a lack of moving parts, and their high efficiency. Their uses are predominantly in the chemical industry, such as for the extraction of toluene from benzoic acid with water as a solvent.

2. *Agitated separation towers*: Agitated separation towers include motors that rotate the disks within the column, hence agitating the liquids within the tower as they make their way through it. Different agitators are used, such as Oldshue-Rushton or Scheibel. The heavier phase enters from the top of the column, where it encounters the lighter phase in the rotating disk, and they are both mixed due to the agitation. They then move to the external stage where they settle and split; mass transfer occurs during this period. Due to its lower density, the heavier phase moves down the column where it is mixed with fresh lighter liquid. This entire cycle is repeated down the entire column. Baffles reduce axial mixing and instead generate many mixing regions where mass transfer can occur. Many modern separators have incorporated a mechanism to agitate the liquids to induce phase dispersion. This type of separator is predominantly used in the oil industry to de-asphalt the oil and also to treat lubricants by removing furfural (Wardle and Weller, 2013).

8.2.3 Applications in Biochemical Engineering

Most of the applications of conventional liquid-liquid separation discussed until now have been either in the chemical or petrochemical industries, but with large advances being made in the biochemical engineering field, liquid-liquid separation is also being applied in the following projects:

1. *Recovery of oil from algae broths*: During the fermentation process, many biofuels, such as methane and alcohols, as well as other chemicals, are produced. The algae used during the fermentation process are recovered and the biofuels are purified using liquid-liquid separation. It is a sustainable process for purifying the products of fermentation because it has a low energy requirement and it is also highly cost-effective. In addition, when hydrolyzed lignocellulosic biomass is fermented, chemicals, such as phenols and long-chain hydrocarbons, that are produced as by-products start to inhibit the growth of the algae that produce the biomass. Therefore, liquid-liquid separation is used to remove these compounds from the fermentation process.
2. *Recovery of carboxylic acids and biofuels from biomass*: Careful pyrolysis of biomass generates bio-oil. The product that is generated at the end contains various long-chain, complex hydrocarbons. This results in the bio-oil having a low viscosity and a high water content, reducing its energy content during combustion. Therefore, liquid-liquid separation is used to separate the components in bio-oil depending on the chemical components present and their polarity. This has a large impact on the quality of bio-oil produced.

3. *Removal of high boiling organics from wastewater*: The latest technologies remove micropollutants from wastewater in order to recycle and reuse it sustainably. One such technology is liquid-liquid separation, which removes phenols, aromatics, and nitrates, all of which are known to cause ill health. Because of the large number of different components present in wastewater, a combination of solvents must be used to remove these organic compounds from it. It would be ineffective to use distillation in this case as the micropollutants have high boiling points, making liquid-liquid separation a more economical solution.
4. *Protein separation and purifications*: Since proteins are prone to denaturation at high temperatures, the most suitable way to extract and purify them from solutions is by using liquid-liquid separations. A variety of factors influence the extraction of the proteins, such as the partition coefficients, the time period of contact between the solvent and the protein solution, the composition of the solvent, mixing of the two phases, and the contact area.
5. *Agricultural chemical extraction*: Wastewater also contains agricultural chemicals, for example, herbicides, pesticides and fertilizers. These can be separated using liquid-liquid separation. Similarly, metal ions and organic compounds can also be removed using the same method.

8.2.4 Advantages and Limitations of Conventional Liquid-Liquid Separation

The advantages of liquid-liquid separation are that it is effective for compounds with high boiling points even in low concentrations. It is effective for heat-sensitive compounds like large organic molecules. It can also generate very large capacities using minimum energy, leading to lower operational costs.

Disadvantages include the large consumption of possibly toxic solvent, which when recycled in the system requires costly equipment. Thus, the appropriate selection of the solvent in achieving a sustainable process creates challenges for large-scale industry implementation of the process.

Most industrial processes seek to produce a product with high purity. Liquid-liquid separation tends to be a long and inefficient process because the yield of product is not as high as it is if other separation methods, such as distillation, are used. Therefore, a secondary extraction method, for instance, distillation, chromatography, or crystallization, may be required. These secondary methods have high operational costs associated with them, as well as large energy loads. Therefore, during the planning stage of a project, all factors should be considered to ensure that the process can achieve the desired yield (Dahal et al., 2016).

However, with new liquid-liquid separation methods, such as LBPS and LBFS, being developed, it seems that the limitations of conventional liquid-liquid separation methods will be overcome. The newer techniques are incorporated into the existing technologies, making them more efficient.

8.3 LIQUID BIPHASIC PARTITIONING SYSTEM (LBPS)

8.3.1 Mechanism of LBPS

LBPS generally consists of polymer/polymer and polymer/salt systems. To form a polymer/salt LBPS, one must dissolve polymer and salt phases in water beyond their critical level (Li et al., 2013), which typically consists of 70%–80% water content. One phase is enriched with solutes, while another phase remains with the parent sample. Operation of LBPS begins with the gentle mixing of fermented broth or parent material containing the solute of interest along with the LBPS. When polymer/polymer or ionic salts are mixed, these phases tend to aggregate and form two phases due to steric exclusion (Iqbal et al., 2016). The mixture is allowed to settle by gravitational force to complete the solute partition. Note that centrifugation can be used to hasten biphasic system formation provided it does not damage any biomolecules that are to be extracted. During settling of aqueous phases, solutes to be extracted are partitioned between the initial phases that were mixed together. Thus, extraction of the solute is from the sample in the top phase. The recovery of solute from the top phase can be done by back extraction and solvent washing.

Considering the driving forces that influence the partition of biomolecules in LBPS, the exact contributing mechanism is still not well understood (Goja et al., 2013), despite many available thermodynamic models. For simplicity, this chapter will address Albertsson's model to highlight the six possible types of partitioning and their respective driving forces. Albertsson's model is one of the commonly used models in describing the behavior of protein partition. The proposed model (Table 8.2) showed the partition effects of the possible mechanisms.

TABLE 8.2
Types of Partition Mechanism and Their Driving Forces, as Described in Albertsson's Model

Partition Mechanism	Description of Driving Force
Biospecific affinity	Partition depends on the affinity of proteins and attached ligands in the phases used.
Conformation-dependent	Biomolecules conformation dictates partition results.
Electrochemical	Extraction by exploiting net charge of particles between electrical potential of biphasic components.
Hydrophobic affinity	Uses hydrophobic properties of biphasic system to separate biomolecules based on their hydrophobicity.
Size-dependent	Molecular size of biomolecules dominates partition.

Source: Grilo, A. L. et al., *Sep. Purif. Rev.*, 45, 68–80, 2014.

Alcohol/salt–based LBPS is usually formed by two immiscible phases consisting of short-chained aliphatic alcohols and usually an inorganic salt. To obtain a more environmentally friendly result, however, organic salts, such as citrates and acetates, can be used to replace the inorganic salt (Lo et al., 2018). Compared to against conventional polymer-based LBPS, the target proteins are easily recovered by removal of alcohol with evaporation, distillation, or crystallization (Lo et al., 2018). Alcohol/salt–based LBPS has several more advantages, including faster segregation of phases, larger capacity in scale up, lower toxicity to the environment, and relatively low cost (Lo et al., 2018). These systems have been used to purify biomolecules; examples include proteins, natural compounds, enzymes, acids, antibiotics, and nucleic acids.

On the other hand, the use of ionic liquids combines both the advantages of LBPS and of the ionic liquids from over 10^6 simple ionic liquids, such as imidazolium and pyridinium ionic liquids (Alvarez-Guerra and Irabien, 2014). Also, the inorganic salts can be replaced with carbohydrates, such as glucose, fructose, or sucrose; however, the use of these carbohydrates lowers the phase forming ability in comparison to inorganic salts (Alvarez-Guerra and Irabien, 2014). Such replacements are often noncharged, biodegradable, nontoxic, and renewable feedstock, which is generally desirable. Ionic-liquid-based LBPS too has the ability to recycle or concentrate hydrophilic ionic liquids from aqueous solutions compared to conventional polymer-based LBPS (Alvarez-Guerra and Irabien, 2014). Dissociation of conventional salts into ions in phase solutions is observed. Therefore, this simplifying the recycling process together. The formation of ionic liquid biphasic system follows the Hofmeister series (Alvarez-Guerra and Irabien, 2014), which is defined by the ability of ions to precipitate proteins. This shows interesting possibilities in overcoming constraints with polymer-based LBPS, such as the lower viscosity, short separation time, and most important the ability to tune-phase polarities. The tuning is usually done by adequately selecting from a vast variety of multiple ionic liquid choices (Lee et al., 2017). As a result, it can be specifically engineered to partition biosolutes of interest.

A thermos separating polymer (TSP) based LBPS usually consists of random, deblock and triblock copolymers of hydrophilic ethylene oxide (EO) and hydrophobic propylene oxide (PO). Hence, they are named EOPO co-polymers. Biomolecule separation can be done by utilizing cloud point extraction, which is a temperature-induced phase separation where an aqueous TSP solution is heated above the respective low critical solution temperature (LCST) or the cloud point (CP) temperature. This particular separation method is due to the unique characteristic of the process that allows separation to occur and makes recycling of the TSP feasible. However, this type of LBPS is undesirable for heat-sensitive biomolecules due to the high temperature utilized during the process.

8.3.2 Variation of LBPS

Different types of LBPS are shown in Table 8.3. Researchers have used different combinations of different phase-forming components to form LBPS for the separation and purification of many biotechnological products.

TABLE 8.3
Different Types of LBPS and Their Efficiency

Type of LBPS	Composition	Product Extract	Results	Reference
Polymer/polymer	PEG/dextran	Serine protease from mango peel	• Overall yield of 97.3% • Overall purification factor of 14.37	Mehrnoush et al. (2012)
Polymer/salt	PEG/KH_2PO_4	Laccase from *Hericium erinaceus*	• Overall yield of 99% • Purification factor of 8.03 ± 0.46	Rajagopalu et al. (2016)
Alcohol/salt	Ethanol and 1-propanol/ $K_3C_6H_5O_7$	Enhanced green fluorescent protein (EGFP) from *E. coli* lysate	• Average EGFP yield of 75.7% • Average purification factor of 11.34	Lo et al. (2018)
Ionic liquid	[Ch][BES]/ Polypropylene glycol (PPG)	Microbial lipase from *Burkholderia cepacia*	• Recovery yield of 99.30 ± 0.03 • Purification factor of 17.96 ± 0.32	Lee et al. (2017)
Thermo-separating polymer	Ethylene oxide-propylene oxide (EOPO)/NaCI	Polyhydroxyalkanoates (PHAs) from *Cupriavidus necator* H16	• Recovery yield of 94.8% • Purification factor of 1.42-fold	Leong et al. (2017)

8.3.3 Influencing Parameters on Partitioning of Biomolecules

8.3.3.1 Tie-Line Length

Generally, LBPS can be described using the phase diagram shown in Figure 8.1, where the *x*-axis represents the bottom phase constituent and the *y*-axis represents the top phase. Any composition of LBPS that lies above the binodal curve concentration (TCB = Binodal curve, C = critical point, TB = Tie line, T = composition of the top phase, B = composition of the bottom phase) results in the formation of the biphasic system. Note that the straight line of TCB is known as the tie-line length. The position of the binodal and tie line of TCB depends on the pH, temperature, molecular weight of polymers, and its ionic strength (Millqvist-Fureby, 2014). Measurement of the tie-line length presents the degree of incompatibility of the biphasic system (Grilo et al., 2014); therefore, it is a useful parameter to determine the amount of constituent polymers from the two phases to achieve this. The greater the tie-line length, the higher the partition coefficient, which in turn increases yield of biomolecules extraction (Goja et al., 2013). With the increasing tie-line length, LBPS becomes more hydrophobic due to the reduction of water content in the phases. In cases dealing with PEG/salt systems, a higher tie-line length promotes the salting out effect, which eventually leads to more protein to partition within the PEG rich phase (Goja et al., 2013). However, protein may precipitate at the biphasic interface if the solubility in the phases is inadequate; this depends on the natural properties of the biomolecules themselves.

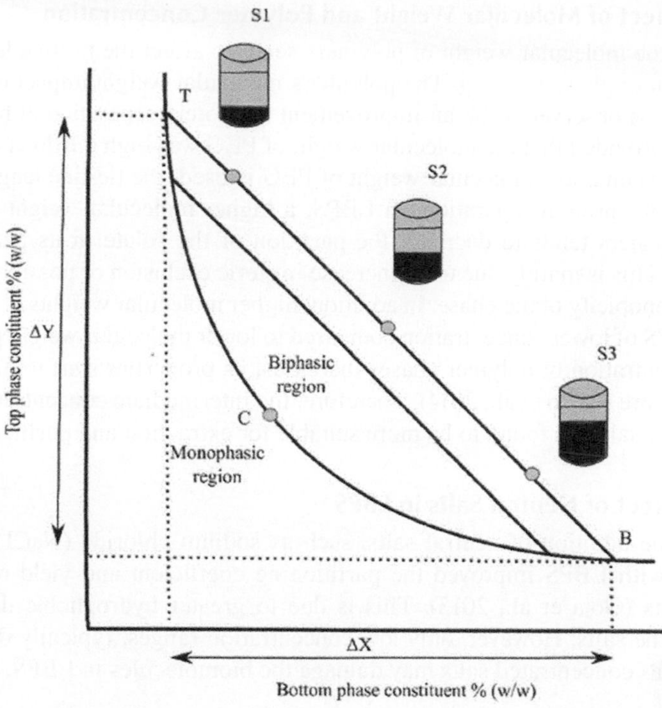

FIGURE 8.1 Schematic diagram of a phase diagram illustrating tie-line length.

8.3.3.2 Temperature

Temperature has a significant impact on the biphasic equilibrium composition, and it is often recommended to exert strict control over the temperature during research. Temperature affects several other physicochemical properties, such as interfacial tension, viscosity, and density indirectly (Iqbal et al., 2016). In general, a lower temperature is preferred in polymer/polymer systems, while polymer/salt systems favor higher temperatures (Grilo et al., 2014).

8.3.3.3 Effect of pH

The pH value of biphasic systems affects the net charge of the biomolecules solute and, as a result, induces an electrochemical difference (Iqbal et al., 2016) in LBPS for separation. The partition of solute in the context of pH depends on the biphasic system's pH relative to the solute's isoelectric point (pI). As an example, bovine serum albumin (BSA) naturally has an isoelectric point of 5.0. Whenever BSA is placed in the LBPS with a pH higher than 5.0, the net charge of BSA turned negative and is attracted to the phosphate-rich phase (Grilo et al., 2014) to counteract this change in the EOPO-phosphate LBPS. In contrast, when the bulk phase is below the BSA isoelectric point, the BSA net charge turned positive and the proteins partition favored the EOPO-rich phase.

8.3.3.4 Effect of Molecular Weight and Polymer Concentration

Changes in the molecular weight of polymers seem to affect the tie-line length and concentration of phase forming. The polymer's molecular weight impact on the tie-line length was observed to be an improvement in protein separation at higher tie-line lengths provided that the molecular weight of PEG was high (Grilo et al., 2014). Conversely, when a low molecular weight of PEG is used, the tie-line length has no significance for protein separation. In LBPS, a higher molecular weight in phase-forming polymers tends to decrease the partition of the solute in its phase (Goja et al., 2013). This is mainly due to an increase in steric exclusion or possibly changes in the hydrophobicity of the phase. In addition, higher molecular weights of polymers form an LBPS of lower concentration compared to lower molecular weight polymers. As the concentration of polymer phases increases, its properties start to differ from water even more (Grilo et al., 2014). Therefore, the intermediate concentration range of the PEG or salt was found to be more suitable for extraction and purification.

8.3.3.5 Effect of Neutral Salts in LBPS

Generally, the addition of neutral salts, such as sodium chloride (NaCI), that are compatible with LBPS improved the partitioning coefficient and yield of biomolecule extracts (Goja et al., 2013). This is due to greater hydrophobic differences induced by the salts. However, only low concentration ranges, typically 0.0–1.0 M, are allowed as concentrated salts may damage the biomolecules in LBPS.

8.3.4 Applications of LBPS

Research has shown great development and application of LBPS in several industrial fields, including protein extraction, enzymes, monoclonal antibodies (mAbs), metal ions, and phytochemicals. These applications are not limited to bioseparation as LBPS has found success, from recovering drug residue from wastewater to environmental remedies. As an example, LBPS was found to be an excellent alternative for the removal of textile dyes (Iqbal et al., 2016) that is more economically feasible and environmentally friendly. Another great example of a nonbiological application was the extraction of cadmium and chromium (Iqbal et al., 2016) using an ionic liquid LBPS. LBPS is also a great analytical tool for understanding the fundamental behavior of protein and its chemical properties because biphasic systems can be fixed to remain relatively constant to carry out the study. Fractionation of biomolecules in microfluidic is also possible by using LBPS, owing to its free label cell separation feature (Gossett et al., 2010).

In addition, LBPS was able to recover drug residues, such as roxithromycin from water, chloramphenicol from shrimp, and ciprofloxacin from eggs, with promising results (Iqbal et al., 2016). It averaged a recovery rate of at least 90%. The application of LBPS on phytochemicals, such as anthocyanins and betalains, was studied, and it was reported that the method did not alter the extracted natural pigments or affect their antioxidant activity (Iqbal et al., 2016) compared to conventional techniques. As for applications in protein and enzyme, some great examples would be the purification of recombinant enhanced green fluorescent protein (EGFP) from *E. coli*

(Lo et al., 2018) and the extraction of microbial lipase from *Burkholderia cepacia* (Lee et al., 2017). LBPS can also separate and purify monoclonal antibodies such as human IgG (Schenke-Layland and Walles, 2013) from CHO supernatant cell cultures.

8.3.5 Advantages of LBPS

LBPS has significant advantages over conventional methods of protein separations, such as chromatography and electrophoresis, due to its relatively simple and huge scalability in equipment design (Iqbal et al., 2016). These reasons are supported by the fact that carrying out protein separation via column chromatography can be tedious in terms of methodology, and it requires very high pressure to drive eluent through packed columns, experiences high pressure drop, has a complicated equipment design, and is very expensive to build, especially for columns that require specifically small pore packing. The process can be carried out only in small amounts of kilograms under batch operations, with its packed column frequency unloaded for biomolecular recovery.

When compared to LBPS, the equipment design and concepts are much simpler and less sophisticated. Most important, it can be scaled up to tonnes instead of kilograms and can be run in continuous operation with ease (Goja et al., 2013), making LBPS an attractive alternative in high-cost protein separations. In addition, LBPS is selective toward specific proteins and biomolecules of interest under suitable phase combinations and operating conditions. Many research studies have reported high recovery and purity of product isolation in such a straightforward process. Although the research is conducted on an experimental basis, there is an enormous potential for application in bioprocesses. Nonetheless, this highlights the strength of LBPS for "capture and purify" stages and considers the two-step process in downstream processing. LBPS also requires very low energy requirements to operate. The features of LBPS are quite favorable for protein separation processes. For example, high water content presents a gentle environment and lower interfacial tension of solute partition (Grilo et al., 2014), while ambient conditions such as room temperature (<35°C) are beneficial for any heat-sensitive biomolecules.

8.3.6 Disadvantages of LBPS

Despite its benefits, LBPS has some limitations and disadvantages. One is the difficulty in isolating extracted solutes from the polymer/top phase and predicting the behavior of targeted proteins in the biphasic system (Goja et al., 2013). This makes the monitoring of product characteristics and quality control difficult. In addition, to fully optimize LBPS, many trial-and-error experimental runs are required to determine the optimal conditions to partition a specific bioproduct, and this requires a considerable amount of time.

Alternatively, formations of LBPS with polymer phases tend to be visually opaque, which at times can be misleading in determining the exact level surface interface. The established hydrophobic and hydrophilic properties in polymer and salt phases pose limited polarity or hydrophobicity differences between the two phases (Freire et al., 2012). Thus, this prevents a wide application of polymer-based LBPS for extraction of biomolecules.

In the context of design and scale-up, most polymer phases tend to be quite viscous. This makes handling the stream difficult, slowing down mass transfer and phase separation as they proceed. A tremendous amount of chemicals is required for phase forming of LBPS, especially polymer-based materials, which incurs greater costs. This has some carryover effects because there is a greater risk for environmental pollution due to elevated salt concentration in the effluent and the recycling of phase-forming polymers (Goja et al., 2013).

8.4 LIQUID BIPHASIC FLOTATION SYSTEM (LBFS)

8.4.1 Principle of LBFS

Liquid biphasic flotation system (LBFS) is a new technology that is a combination of LBPS and solvent sublation (SS). LBFSs integrate the principle of LBPS and the mass transfer mode of SS, which involves bubble adsorption. SS applies the mechanism of effective adsorption between the surface of bubbles and surface-active material interaction in aqueous phase. The rising bubbles carry surface-active materials that release them at the organic solvent phase, which is at the top column (Lee et al., 2016; Mathiazakan et al., 2016). Typically, a biphasic system consists of alcohol and salt solution, polymer and salt solution, or polymer and polymer solution. For lower operational cost, an alcohol/salt solution system is favored, and alcohol is easier to remove by evaporation. In LBFS, a gas stream is introduced directly into the equipment that flows in an ascending manner. According to SS mass transfer mode, the gas stream acts as carrier medium and thus allows surface active components identified as hydrophobic in nature from the crude feed to attach themselves on the air bubbles' surface in the salt solution. Thus, bubbles carrying the surface active components travel to the top phase that contains alcohol, which is hydrophilic in nature, and dissolves into the solution and accumulates (Phong et al., 2017). The ascending gas bubbles create the flotation effect from the aqueous phase to the organic solvent phase. The gas stream is normally inert gaseous to prevent unnecessary chemical reaction. The accumulated surface-active material at the top phase is recoverd by the evaporation method.

8.4.2 Typical Setup of LBFS

A typical lab-scale LBFS consists of a glass column equipped with a sintered disk of G4 porosity at the bottom of the column (Figure 8.2). The sintered disk of G4 porosity is used to ensure that bubbles of even size enter the column. The sintered disk is attached to a compressed air system that supplies gas and generates bubbles to the column. A flow meter monitors the flow rate of gas entering the glass columns. The majority of studies on LBFS were conducted on a lab scale; however, a pilot scale of LBF was introduced in 2016 by Mathiazakan et al. that used a water tank made of HDPE prepared with double-woven fabric of G4 porosity layered at the bottom of the tank.

LBFS is prepared by mixing the crude extract with the aqueous phase, and then the solution that is the bottom phase is transferred into the flotation column. Subsequently, a layer of aqueous polymer or organic solvent is added into the column as the top phase. The solutions must be set up to create immiscible phases. Gas bubbles are

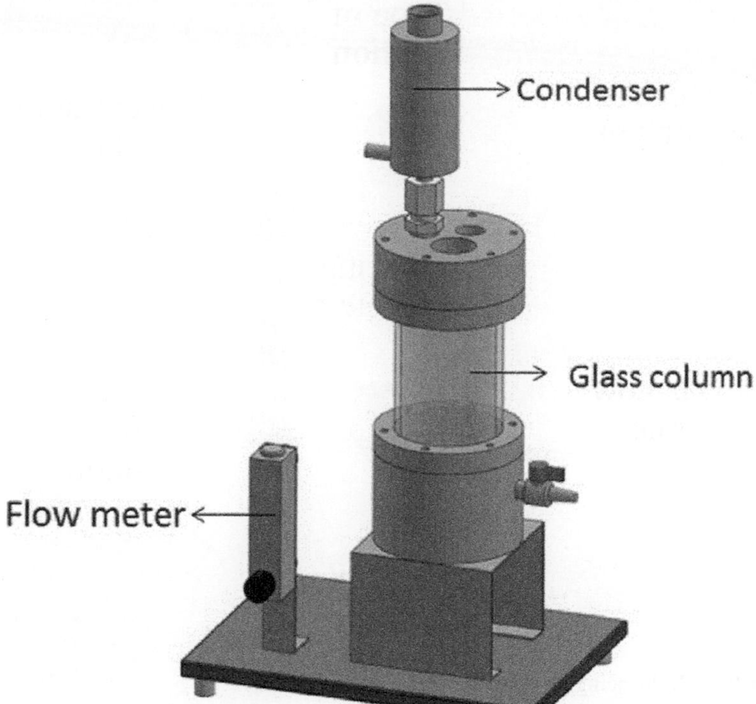

FIGURE 8.2 LBFS equipment. (Obtained from Sankaran, R. et al., *J. Clean. Prod.*, 184, 938–948, 2018.)

bubbled into the bottom phase containing the aqueous salt solution and crude feedstock by a sintered glass disk at the bottom of flotation cell. Finally, the bubbles rise to the top phase where the mass transfer process takes place by SS mechanism.

8.4.3 Optimization of Parameters of LBFS

The parameters of LBFS that affect separation efficiency can be seen in Figure 8.3. These parameters will be discussed in following subsections.

8.4.3.1 Type of Solution

LBFS consists predominantly a polymer and aqueous phase system (Bi et al., 2009, 2011, 2013; Chang et al., 2014; Li and Dong, 2010; Lin et al., 2015; Md Sidek et al., 2016). PEG is commonly used in studies as it has showed good separation efficiency; it also has the lowest environmental risk and is more economically feasible compared to other polymers. But to achieve a sustainable approach, substituting a hydrophilic organic solvent, alcohol, for the polymer is seen as a promising method. Isolation of biomolecules from the alcohol phase is easier because it involves a simple evaporation method. Alcohol has lower viscosity, is cheaper, and is environmentally friendly

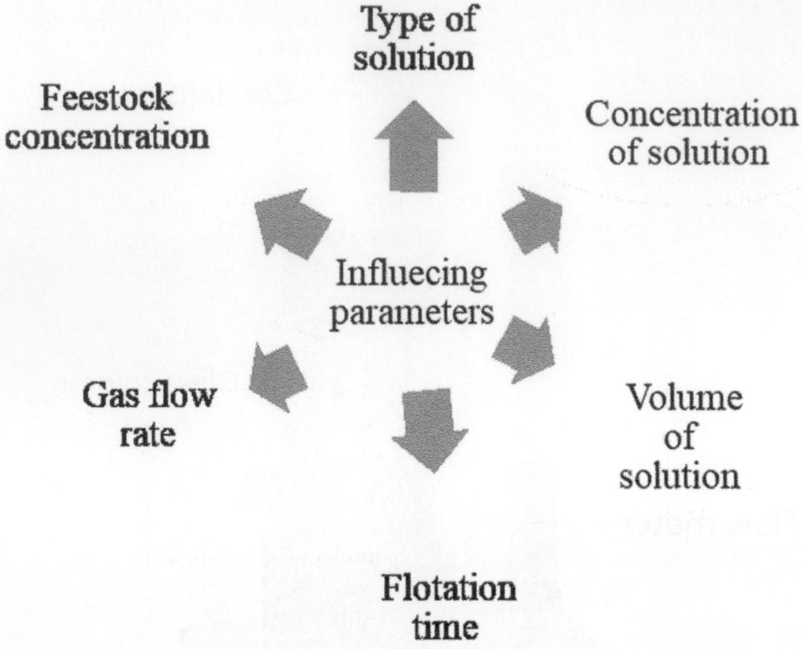

FIGURE 8.3 Influencing parameters of LBFS.

compared to polymer. In 2013, Show et al. introduced the application of hydrophilic organic solvent and inorganic salt in LBFS where lipase was recovered from *Burkholderia cepacia* ST8. An efficiency rate of 90.2%, a concentration coefficient of 16.1%, and 99.2% yield of lipase were achieved in a 2-propanol/potassium phosphate system (Show et al., 2013).

Based on a review of the literature, the typical alcohols used are methanol, ethanol, 1-propanol, and 2-propanol. A suitable alcohol is an aliphatic compound that has high hydrophobicity, a low evaporation rate, and low molecular weight. Higher hydrophobicity increases the effectiveness of gas bubbles containing surface-active materials to dissolve in alcohol-rich phase, and lower evaporation rate prevents the alcohol from drying out during the separation process. Normally, 1-propanol is identified as the suitable alcohol compared to methanol, ethanol, and 2-propanol according to their properties; 1-propanol used in Phong et al. (2017) showed an optimal result. However, ethanol was seen as the suitable alcohol when combined with K_2HPO_4 in betacyanins extraction from the peel and flesh of *Hylocereus polyrhizus*. Besides, a 1- or 2-propanol combination with salt shows negative separation efficiency despite their high level of hydrophobicity (Leong et al., 2018).

Next, typical types of salts used in experiments are potassium dihydrogen phosphate (KH_2PO_4), dipotassium hydrogen phosphate (K_2HPO_4), magnesium sulphate ($MgSO_4$), and ammonium sulphate ($(NH_4)_2SO_4$). For protein extraction situations, salts with lower molecular weight have a greater tendency to separate with increasing partition coefficient. Salts that show salting out effect have better biphasic system formation and separation efficiency, and this can be achieved by manipulating the concentration of salt. Overall, there is no certain alcohol/salt combination because each biomolecule possesses individual traits and properties. Thus, it is recommended to conduct several tests to select the appropriate solvent before executing any experiments.

8.4.3.2 Concentration of Solution

Concentration of aqueous phase in which the salt solution stimulates the salting-out effect creates an immiscible two-phase system. Hence, higher concentration of aqueous phase promotes higher salting-out effect. Studies showed that separation efficiency increased with increasing concentration of aqueous phase. Bi et al. (2009) studied a range between 250 and 400 g/L of $(NH_4)_2SO_4$, and concentration above 350 g/L and demonstrated a higher separation efficiency and distribution ratio of penicillin G.

8.4.3.3 Volume of Solution

In LBFS, a small dosage of polymer is sufficient because larger doses of polymer have an equivalent effect on the separation efficiency, as reported by Bi et al. (2009), which is an added advantage for this method. Even though small doses of polymer are required in LBFS, isolation of desired biomolecules is still seen as a more complex extraction method compared to recovery in alcohol solution.

A moderate volume of salt and alcohol is substantial to obtain maximum separation efficiency. A lower alcohol volume promotes a high evaporation rate, which reduces interaction between surface-active material and alcohol particles. A biphasic system is difficult to create. In addition, a lower salt solution volume causes high viscosity of the crude feedstock solution. When the crude feedstock solution is too viscous, the formation of rising bubbles is disrupted. Separation efficiency increases with increasing volumes of alcohol and salt; however, it is not economically advisable as higher volume increases the operational cost. An optimum moderate volume ratio of alcohol to salt solution should be 1:1. In betacyanins extraction from the peel and flesh of red-purple pitaya, 1:2 volume ratio is found to obtain the highest separation efficiency (Leong et al., 2018). Further increase in volume does not affect the extraction; thus, it is unnecessary.

8.4.3.4 Flotation Time

Flotation time does not affect separation efficiency but it does affect recovery yield. A longer flotation time increases the limitation of hydrophobic interaction in an alcohol-rich top phase. In the recovery of protein from wet microalgae, 10 minutes from the range of 5–25 minutes is sufficient where 94.07% of protein is recovered instead of 93.65% at 25 minutes (Phong et al., 2017). Corresponding to the results obtained from lipase recovery from *Burkholderia cepacia* by Show et al. (2011), moderate flotation time in LBF system is adequate.

8.4.3.5 Gas Flow Rate

The flow rate of gas bubbles entering the column affects the effectiveness of targeted biomolecules on the surface of the bubbles and the velocity of mass transfer of biomolecules between bottom and top phases. The higher the gas flow rate, the higher the accumulation rate of biomolecules at the top phase, and that results in limited space for further dissolution of gas bubbles and creates foam. In the recovery of bacteriocin-like inhibitory substance (BLIS) from *Pediococcus acidilactici* Kp10, a gas flow rate ranging from 10–50 mL/min shows a better separation efficiency at 20 mL/min (Md Sidek et al., 2016). Besides, nitrogen gas flow rate of 40 L/min from the range of 30–60 L/min is chosen as the optimum flow rate in separation of penicillin G (Bi et al., 2009).

8.4.3.6 Crude Concentration

The equilibrium and composition of a biphasic system is affected with a high amount of crude feedstock. Thus, a lower recovery yield and separation efficiency is obtained (Leong et al., 2018).

8.4.4 Recycling of Top- and Bottom-Phase Components

In order to increase the advantages of LBFS and meet sustainability standards, recycling was introduced to minimize the operational cost and environmental impact.

In the extraction of lipase from *Burkholderia cepacia* ST8, EOPO/$(NH_4)_2SO_4$ LBFS is incorporated with recycling technique (Show et al., 2011). Lipase was extracted directly from the culture broth in a single recycling step. The recycling approach in LBFS was conducted in two-step purification, that is, in a primary and secondary system. In the primary system, the desired protein was recovered from the EOPO phase, whereas in the secondary system, EOPO polymer phase was recovered from the primary system. The secondary system was heated to above the LCST. Then, the EOPO phase was removed from the primary extraction and allowed to sit in a water bath at a temperature of 65°C for 15 minutes. The concentrated EOPO copolymer from the secondary system was recycled for the following primary LBFS. After the recycling process, 98.22% of protein was recovered (Show et al., 2011).

Because of unfavorable methods and limitations, such as high cost and slow separation process, faced in the recovering polymer phase, a recent study on the same process, lipase recovery with the replacement of EOPO to 1-propanol, was conducted. The recycle LBFS consisting of 1-propanol and ammonium sulphate was scaled up with the consideration of three optimizing parameters, which are volume and concentration of crude lipase, type of alcohol and salt, and concentration of alcohol and salt (Sankaran et al., 2018).

Instead of five successive recycling steps of the bottom aqueous phase in Show et al. (2011), eleven successive recycling steps of the bottom aqueous phase were conducted by Sankaran et al. (2018). These authors demonstrated a lipase recovery yield of above 50%. Recovery of alcohol was conducted in a simple process via vacuum rotary evaporation. Hence, recycling of alcohol/salt LBFS has advantages, including ease of phase component recovery, a higher number of recycling steps, lower cost, and environmental safety.

8.5 COMPARISON OF CONVENTIONAL LIQUID-LIQUID SEPARATION, LBPS, AND LBFS

Different types of liquid-liquid separation, that is, conventional liquid-liquid separation, LBPS and LBFS, are summarized in Table 8.4. Their principles, applications, advantages, and limitations are discussed in the table.

TABLE 8.4
Summary of Liquid-Liquid Separation Techniques

	Conventional Liquid-Liquid Separation	Liquid Biphasic Partitioning System (LBPS)	Liquid Biphasic Flotation System (LBFS)
Principle	• Bringing two immiscible or liquids in a countercurrent manner • Mechanism: mass transfer of the solute from the mixture to the solvent	• Dissolve both phases in water above critical threshold to form aqueous (biphasic system) • Gentle mix • Gravity settling/centrifugation • Partitioning of solute of interest reached equilibrium	• Integration of LBPS and solvent sublation (SS) • Addition of inert gas into the system, which acts as a carrier; the inert gas is bubbled upward from the bottom of the column through the bottom phase and followed by the top phase
Applications	• A key operation in a wide range of industries, such as biochemical, metallurgical, pharmaceutical, petrochemical, food processing, and nuclear industries globally • Various types of technologies developed: mixer-settler, packed separation tower, centrifugal separator, and so on	• Analytical tool for behavior and properties of proteins, quantify and qualify as well as label-free cell separation • Waste recovery: drug residue • Pharmaceutical proteins, enzymes and mAbs recovery: human IgG, lipase, penicillin	• Various LBFS have been applied successfully to deal with surface-active compounds in bioseparation and bioengineering, for example, lipase, penicillin G, baicalin, chloramphenicol, and so on • Pharmaceutical industry: antibiotics recovery • Agricultural industry: extraction of biocide (*ortho*-Phenylphenol [OPP]) used as preservatives • Nutrient extraction

(*Continued*)

TABLE 8.4 (*Continued*)
Summary of Liquid-Liquid Separation Techniques

	Conventional Liquid-Liquid Separation	Liquid Biphasic Partitioning System (LBPS)	Liquid Biphasic Flotation System (LBFS)
Advantages	• Highly effective toward compounds with high boiling points • Effective toward heat-sensitive compounds like large organic molecules • Can also generate very large capacities, using a minimum of energy consumption, leading to lower operational costs	• Low energy requirements • Simple, selective, and scalable system • High yield, recovery, and purity of biomolecules • Two-step bioprocessing (extraction + purification) • Biocompatible media and low interfacial tension • Relatively lower investment	• High enrichment factor • Biofriendly • Low cost • Low environmental impact • Easy recovery of the phase-forming chemicals • Simple operation
Limitations	• A secondary extraction method, for instance, distillation, chromatography, or crystallization, may be required • Increase economical costs • Large consumption of possibly toxic solvent	• Gaps in knowledge: no exact science behind partitioning mechanism yet and optimization for specific bioproducts requires trial and error • Design and scale-up: most used polymers are usually viscous, requires lots of chemical for phase forming, slows phase separation/back extraction; some scale-up did not achieve similar partitioning results • High risk of contamination and reduced separation efficiency	• Unknown appropriate parameters for larger scale

8.6 CONCLUSION

Conventional liquid-liquid separation methods are effective separating components with high boiling points even in low concentrations, but their products have low purification. As a consequence, they require a secondary extraction method, for instance, distillation, chromatography, or crystallization. On the other hand, LBPS has high separation efficiency, but the method still possesses challenges, such as enormous chemical requirements, environmental risks, and long separation process. Hence, with the integration of LBPS and SS, LBFS demonstrates a better potential in the liquid-liquid separation with convincing separation efficiency and product recovery yield. In addition, alcohol/salt–based biphasic systems present a more sustainable approach and should be studied further.

REFERENCES

Alvarez-Guerra, E., and Irabien, A. 2014. "Separation of proteins by ionic liquid-based three-phase partitioning." In *Ionic Liquids in Separation Technology*. Amsterdam, the Netherlands: Elsevier.

Bi, P.-Y., Chang, L., and Dong, H.-R. 2011. "Separation behavior of penicillin in aqueous two-phase flotation." *Chinese Journal of Analytical Chemistry*, 39: 425–428.

Bi, P.-Y., Chang, L., Mu, Y.-L., Liu, J.-Y., Wu, Y., Geng, X., and Wei, Y. 2013. "Separation and concentration of baicalin from *Scutellaria baicalensis* Georgi extract by aqueous two-phase flotation." *Separation and Purification Technology*, 116: 454–457.

Bi, P.-Y., Li, D.-Q., and Dong, H.-R. 2009. "A novel technique for the separation and concentration of penicillin G from fermentation broth: Aqueous two-phase flotation." *Separation and Purification Technology*, 69: 205–209.

Chang, L., Wei, Y., Bi, P.-Y., and Shao, Q. 2014. "Recovery of liquiritin and glycyrrhizic acid from *Glycyrrhiza uralensis* Fisch by aqueous two-phase flotation and multi-stage preparative high performance liquid chromatography." *Separation and Purification Technology*, 134: 204–209.

Cleveland, C. J., and Morris, C. 2014. "Mixer-settler." In *Dictionary of Energy* (2nd ed.). New York: Elsevier.

Couper, J. R., Penney, W. R., and Fair, J. R. 2012. "Packed towers." In *Chemical Process Equipment—Selection and Design* (3rd ed.). Oxford, UK: Butterworth-Heinemann.

de Haan, A. B., and Bosch, H. 2013. "Centrifugal extractors." In *Industrial Separation Processes—Fundamentals*. Berlin, Germany: De Gruyter.

Freire, M. G., Claudio, A. F., Araujo, J. M., Coutinho, J. A., Marrucho, I. M., Canongia Lopes, J. N., and Rebelo, L. P. 2012. "Aqueous biphasic systems: A boost brought about by using ionic liquids." *Chemical Society Reviews*, 41: 4966–4995.

Goja, A. M., Yang, H., Cui, M., and Li, C. 2013. "Aqueous two-phase extraction advances for bioseparation." *Journal of Bioprocessing & Biotechniques*, 4(1): 1–8.

Gossett, D. R., Weaver, W. M., Mach, A. J., Hur, S. C., Tse, H. T., Lee, W., Amini, H., and Di Carlo, D. 2010. "Label-free cell separation and sorting in microfluidic systems." *Analytical and Bioanalytical Chemistry*, 397: 3249–3267.

Grandison, A. S., and Lewis, M. J. 1996. "Liquid-liquid extraction: Introduction." In *Separation Processes in the Food and Biotechnology Industries—Principles and Applications* (1st ed.) Cambridge, UK: Woodhead Publishing.

Grilo, A. L., Raquel Aires-Barros, M., and Azevedo, A. M. 2014. "Partitioning in aqueous two-phase systems: Fundamentals, applications and trends." *Separation & Purification Reviews*, 45: 68–80.

Harker, J. H., Backhurst, J. R., and Richardson, J. F. 2013. "Centrifugal separators." In *Chemical Engineering (Volume 2)—Particle Technology and Separation Processes* (5th ed.). Oxford, UK: Butterworth-Heinemann.

Iqbal, M., Tao, Y., Xie, S., Zhu, Y., Chen, D., Wang, X., Huang, L. et al. 2016. "Aqueous two-phase system (ATPS): An overview and advances in its applications." *Biological Procedures Online*, 18: 18.

Lee, S. Y., Khoiroh, I., Ling, T. C., and Show, P. L. 2016. "Aqueous two-phase flotation for the recovery of biomolecules." *Separation & Purification Reviews*, 45: 81–92.

Lee, S. Y., Khoiroh, I., Ling, T. C., and Show, P. L. 2017. "Enhanced recovery of lipase derived from *Burkholderia cepacia* from fermentation broth using recyclable ionic liquid/polymer-based aqueous two-phase systems." *Separation and Purification Technology*, 179: 152–160.

Leffler, W. L. 2008. "Applications." In *Petroleum Refining in Nontechnical Language* (4th ed.). Tulsa, OK: PennWell.

Leong, H. Y., Ooi, C. W., Law, C. L., Julkifle, A. L., Ling, T. C., and Show, P. L. 2018. "Application of liquid biphasic flotation for betacyanins extraction from peel and flesh of *Hylocereus polyrhizus* and antioxidant activity evaluation." *Separation and Purification Technology*, 201: 156–166.

Leong, Y. K., Lan, J. C., Loh, H. S., Ling, T. C., Ooi, C. W., and Show, P. L. 2017. "Cloud-point extraction of green-polymers from *Cupriavidus necator* lysate using thermoseparating-based aqueous two-phase extraction." *The Journal of Bioscience and Bioengineering*, 123: 370–375.

Li, M., and Dong, H.-R. 2010. "The investigation on the aqueous two-phase floatation of lincomycin." *Separation and Purification Technology*, 73: 208–212.

Li, Y., Liu, Q., Zhang, M., and Su, H. 2013. "Liquid–liquid equilibrium of the [4-MBP][BF4]–NaCl–H_2O systems at T = 293.15, 303.15, 313.15, and 323.15 K: Experimentation and correlation." *Thermochimica Acta*, 565: 234–240.

Lin, Y. K., Show, P. L., Yap, Y. J., Tan, C. P., Ng, E.-P., Ariff, A. B., Mahammad Annuar, M. S. B., and Ling, T. C. 2015. "Direct recovery of cyclodextringlycosyltransferase from *Bacillus cereus* using aqueous two-phase flotation." *Journal of Bioscience and Bioengineering*, 120: 684–689.

Lo, S. C., Ramanan, R. N., Tey, B. T., Tan, W. S., Show, P. L., Ling, T. C., and Ooi, C. W. 2018. "Purification of the recombinant enhanced green fluorescent protein from *Escherichia coli* using alcohol + salt aqueous two-phase systems." *Separation and Purification Technology*, 192: 130–139.

Mathiazakan, P., Shing, S. Y., Ying, S. S., Kek, H. K., Tang, M. S. Y., Show, P. L., Ooi, C.-W., and Ling, T. C. 2016. "Pilot-scale aqueous two-phase floatation for direct recovery of lipase derived from *Burkholderia cepacia* strain ST8." *Separation and Purification Technology*, 171: 206–213.

Mchugh, M. A., and Krukonis, V. J. 2013. "Coffee decaffeination." In *Supercritical Fluid Extraction* (2nd ed.). Boston, MA: Butterworth-Heinemann.

Md Sidek, N. L., Tan, J. S., Abbasiliasi, S., Wong, F. W. F., Mustafa, S., and Ariff, A. B. 2016. "Aqueous two-phase flotation for primary recovery of bacteriocin-like inhibitory substance (BLIS) from *Pediococcus acidilactici* Kp10." *Journal of Chromatography B*, 1027: 81–87.

Mehrnoush, A., Mustafa, S., Sarker, M. Z., and Yazid, A. M. 2012. "Optimization of serine protease purification from mango (*Mangifera indica* cv. Chokanan) peel in polyethylene glycol/dextran aqueous two phase system." *The International Journal of Molecular Sciences*, 13: 3636–3649.

Millqvist-Fureby, A. 2014. "Aqueous two-phase systems for microencapsulation in food applications." In *Microencapsulation in the Food Industry*. San Diego, CA: Academic Press.

Phong, W. N., Show, P. L., Teh, W. H., Teh, T. X., Lim, H. M. Y., Nazri, N. S. B., Tan, C. H., Chang, J.-S., and Ling, T. C. 2017. "Proteins recovery from wet microalgae using liquid biphasic flotation (LBF)." *Bioresource Technology*, 244: 1329–1336.

Qureshi, N., Hodge, D. B., and Vertes, A. A. 2014. "Use of other separation techniques." In *Biorefineries—Integrated Biochemical Processes for Liquid Biofuels* (1st ed.). Amsterdam, the Netherlands: Elsevier.

Rajagopalu, D., Show, P. L., Tan, Y. S., Muniandy, S., Sabaratnam, V., and Ling, T. C. 2016. "Recovery of laccase from processed *Hericium erinaceus* (Bull:Fr) Pers. fruiting bodies in aqueous two-phase system." *The Journal of Bioscience and Bioengineering*, 122: 301–306.

Dahal, R., Moriam, K., and Seppala, P. 2016. "Downstream process: Liquid-Liquid extraction." Chemical Technology, AALTO University.

Sankaran, R., Show, P. L., Yap, Y. J., Lam, H. L., Ling, T. C., Pan, G.-T., and Yang, T. C. K. 2018. "Sustainable approach in recycling of phase components of large scale aqueous two-phase flotation for lipase recovery." *Journal of Cleaner Production*, 184: 938–948.

Schenke-Layland, K., and Walles, H. 2013. "Strategies in tissue engineering and regenerative medicine." *Biotechnology Journal*, 8: 278–279.

Show, P. L., Ooi, C. W., Anuar, M. S., Ariff, A., Yusof, Y. A., Chen, S. K., Anuar, M. S. M., and Ling, T. C. 2013. "Recovery of lipase derived from *Burkholderia cenocepacia* ST8 using sustainable aqueous two-phase flotation composed of recycling hydrophilic organic solvent and inorganic salt." *Separation and Purification Technology*, 110: 112–118.

Show, P. L., Tan, C. P., Anuar, M. S., Ariff, A., Yusof, Y. A., Chen, S. K., and Ling, T. C. 2011. "Direct recovery of lipase derived from *Burkholderia cepacia* in recycling aqueous two-phase flotation." *Separation and Purification Technology*, 80: 577–584.

Towler, G., and Sinnott, R. K. 2012. "Random packing." In *Chemical Engineering Design—Principles, Practice and Economics of Plant and Process Design* (2nd ed.). Amsterdam, the Netherlands: Butterworth-Heinemann.

Wardle, K. E., and Weller, H. G. 2013. "Hybrid multiphase CFD solver for coupled dispersed/segregated flows in liquid-liquid extraction." *International Journal of Chemical Engineering*, 2013: 1–13.

9 Drying

Chung Hong Tan, Zahra Motavasel, Navin Raj Vijiaretnam, and Pau Loke Show

9.1 INTRODUCTION

Drying refers to the removal of water or solvent from solids, liquids, or semisolids by evaporation, and it is one of the oldest and most important methods for food preservation. Lack of a suitable and timely food-processing technology results in food waste. Postharvest loss in African countries has been estimated at 20%–40%, a significant amount because some parts of Africa showed low agricultural productivity (Abass et al., 2014). Based on a report by the World Bank, the loss of food grains in sub-Saharan Africa (SSH) regions equates to a loss of USD 4 billion every year (Zorya et al., 2011). These losses heavily influence the livelihood of small landholding farmers and are caused by lack of knowledge, poor infrastructure, and difficult or no access to proper drying and storage technologies (Kumar and Kalita, 2017). Drying of food is a common preservation method because it involves the reduction of free unbound water in food to a level where microbial growth is inhibited. Although most drying applications are in the food industry, drying is also an important process in any industry that requires the elimination or minimization of solvents from the final products. These industries include metalwork, woodwork, pulp and paper, chemical, medicine, textile, cosmetics, fertilizer, wastewater treatment, and many others.

In conventional processes, drying is typically the last stage; it involves the removal of water and any volatile solvents from the product while ensuring minimum loss to product quality. For the drying of food, the two major moisture removal methods are drying to produce a solid product, and evaporation to produce a more concentrated liquid. In terms of energy input, drying requires less energy than freezing or canning, and it requires less storage space compared to freeze containers and canning jars. However, the drying process has disadvantages that are common for processes involving heat transfer. The main issues include structural damage to the food product, change in physical appearance, as well as difficulty in regaining the original properties of the food product after rehydration. Various parameters affect overall drying performance such as drying temperature profile, moisture content, and dimensions of the product to be dried. In the food industry, the structure of the food product, the mechanism of water migration, the drying method, and the drying medium are among the key parameters to ensure optimal industrial food dehydration. The various advantages and disadvantages of drying are listed in Table 9.1.

TABLE 9.1
Advantages and Disadvantages of Drying

Advantages	Disadvantages
1. Enables easier handling of materials through the reduction of bulk weight by removing moisture	1. Drying may alter the structure and appearance of food products, making it hard to retain the original properties of the products upon rehydration
2. Minimizes transport and packaging cost of dried materials	2. Sometimes drying costs more compared to canning and freezing
3. Increases shelf life of food products	3. Selecting an appropriate drying method can be time-consuming because the drying method and operating conditions affect the quality and cost of the products
4. Retains the nutritional and physical properties of food products	4. The capital cost and operation cost of drying equipment can be expensive
5. Dried food products are available even during nonharvest season	5. There are drying techniques that may pose health risks to workers such as microwaves and ultrasound
6. Allows early harvesting of crops, which reduces instances of loss due to shattering or consumption by animals and insects	6. Fine powder released in the drying process may lead to lung complications
7. The cheapest mode of preservation of fruits and vegetables	7. Majority of drying systems do not reach their expected drying performance due to low energy efficiency of the process
8. Drying is the main process in baking of food products	8. The theory of the drying process is not always applicable to drying processes in industry
9. In sludge processing, the sludge reduces in volume, which makes it easy to store, transport, package, and sell	
10. In the medical field, drying allows for preservation of platelets for future use	
11. Drying creates timber products such as paper and flax	
12. Essential in the manufacture of detergent powder	

Source: Smith, P. G., An introduction to food process engineering, in *Introduction to Food Process Engineering*, Springer, New York, pp. 1–3, 2011.

This chapter will outline various objectives that are fundamental to the drying process. Fundamentals of drying and the parameters to consider when calculating the drying rate and dryer size will also be discussed in this chapter. The advantages and disadvantages of various drying process, followed by the industrial applications of drying will be introduced. Last, the current drying techniques used in industry and future developments to overcome the disadvantages of the current technology will be reviewed.

9.2 FUNDAMENTALS OF DRYING

Being a physicochemical and thermophysical operation that allows for removal of excess moisture, water activity in dried materials is fundamentally reduced. Before drying, it is crucial to comprehend the structure of the bulk material, the quality of the hot air utilized, the method for moisture migration, the dryer used, and the drying method (Prajapati et al., 2011). Inadequate drying is especially detrimental to the preservation of food and final products with long intended shelf-life because product quality deterioration means diminished profit margin. This implies that selecting a proper dryer and drying technique are the most vital elements for achieving a commercially viable and feasible product.

The principle of heat and mass transfer governs the drying process. Wet material is heated to an appropriate temperature so that the moisture or solvent then vaporizes and diffuses into the environment. The removal of moisture is done by either vaporizing moisture from the material's surface or by encouraging moisture transfer from the inner regions of food to its surface by diffusion, vapor pressure gradient, and cell contraction.

Drying processes can be categorized according to the mode of heat input employed in the dryer, including convection, conduction, infrared radiation, radio frequency, microwaves, or a combination of them (Zhang et al., 2017). The convective method is a direct method where the drying medium (hot air or gas) directly contacts the bulk material and carries the evaporated moisture away from the drying chamber. Over 85% of industrial dryers utilize this method (Wojdyło et al., 2014). In contrast, indirect drying does not involve any contact between the bulk material and the drying medium. Indirect drying utilizes conduction, radiation, radio frequency, and microwaves to alter the final temperature of the material (Vega et al., 2016). Microwave or radio frequency drying allows the electromagnetic energy to be absorbed selectively by water, which is referred to as volumetric heating (Zhang et al., 2016). Typically, less than 50% of the total heat supplied in most direct dryers is used for evaporation. In most instances, regardless of the drying method used, water is the most common solvent that is removed in dryers.

9.3 BASIC TERMINOLOGY

Moisture in materials play a vital role in either maintaining or affecting the materials' quality. Information about the moisture content is vital for safe storage and usage of materials, for instance, food products. Moisture content is described as the quantity of water present in the bulk material and is normally expressed as a percentage. Wet materials can be designated as binary mixtures of a dry solid material and water. The water concentration is expressed as the relative mass fraction of the liquid (X_A), which is simplified by the equation below:

$$X_A = \frac{m_A}{m_C} \tag{9.1}$$

where m_A is the mass of water, m_C is the mass of dry material.

The equilibrium moisture content depends on air conditions as well as the drying properties of the material. In wet basis, the moisture content (MC_{wb}) is defined as a ratio of the weight of water in a material to the total weight of the material, which is expressed by the following equations:

$$MC_{wb}(\%) = \frac{W_w}{W_p} \times 100 \tag{9.2}$$

$$MC_{wb}(\%) = \frac{W_w}{W_w + W_d} \times 100 \quad (9.3)$$

where W_w is the weight of water in material, W_d is the weight of dry matter in material, W_p is the total weight of material.

In contrast, the dry basis moisture content (MC_{db}) is defined as a ratio of the weight of water in a material to that of dry matter, as expressed in the equation below:

$$MC_{db}(\%) = \frac{W_w}{W_d} \times 100 \quad (9.4)$$

The relationship between MC_{wb} and MC_{db} can be defined by rewriting Equation (9.4) as follows:

$$MC_{db} = \frac{W_w}{W_p - W_w} \quad (9.5)$$

$$MC_{db} = \frac{\frac{W_w}{W_p}}{1 - \frac{W_w}{W_p}} \quad (9.6)$$

$$MC_{db} = \frac{MC_{wb}}{1 - MC_{wb}} \quad (9.7)$$

Water in food materials is present in three distinct forms: unbound, bound, and free. Unbound water is the moisture in materials that exhibits vapor pressure equal to the saturated vapor pressure of pure water under the same temperature. Bound water refers to the moisture that applies vapor pressure less than that of pure water under the same temperature owing to moisture retention in minor pores and solutions present in cell walls. In contrast, free water refers to the moisture content that is more than the equilibrium moisture content. Free water can be eliminated with ease by drying under specific conditions of humidity and temperature based on the materials to be dried. In essence, free water encapsulates both unbound water and a portion of bound water (which is part of the larger capillaries in cells) (Kaur et al., 2015).

9.4 PSYCHROMETRIC CONSIDERATION

Atmospheric air, a mixture of water vapor and dry air, is utilized as a medium in various unit operations including mixing, heating, cooling, and transport applications. The thermodynamic characteristics of moist air are usually referred to as psychrometric properties, and the study of the properties of dry air and water vapor mixture is referred to as psychrometry.

9.4.1 Dry Bulb, Wet Bulb, and Dew Point Temperature

The dry bulb, wet bulb, and dew point temperatures are necessary to ascertain the condition of humid air. The dry bulb temperature (T_{db}), commonly called air temperature, refers to the temperature of air measured using a thermometer that is unaffected by the moisture in the air. The wet bulb temperature (T_{wb}) is the adiabatic saturation temperature, and it can be measured using a thermometer with its bulb wrapped in wet muslin. The difference between T_{db} and T_{wb} is called the wet depression. The value of T_{wb} is always between T_{db} and dew point temperature (T_{dp}). The dew point temperature (T_{dp}) refers to the maximum temperature at which air becomes completely saturated with water vapor. Further cooling below this temperature results in the condensation of moisture present in the air. T_{dp} is always lower than T_{db}, but they are the same at 100% relative humidity (Engineering Toolbox, 2004).

9.4.2 Absolute Humidity

The absolute humidity (H) is defined as the mass of water vapor associated with a unit mass of dry air. Even though it is a dimensionless quantity, the units are often quoted as kilogram of water vapor per kilogram of dry air, which can be obtained using the equations below (Beltrán-Prieto et al., 2016):

$$H = \frac{\frac{M_w p_w}{RT}}{\frac{M_a p_a}{RT}} \tag{9.8}$$

$$H = \frac{M_w p_w}{M_a \left(1 - p_w\right)} \tag{9.9}$$

$$H = 0.622 \frac{p_w}{1 - p_w} \tag{9.10}$$

where p_w is the partial pressure of water vapor, kPa; p_a is the partial pressure of dry air, kPa; M_w is the molecular weight of water = 18 g/mol; M_a is the molecular weight of air = 29 g/mol; R is the gas constant = 8.314 J/K/mol; T is the temperature, K.

9.4.3 Relative Humidity

The relative humidity (RH) refers to the ratio of mass of water vapor in moist air (m_v) to the total mass of water vapor needed to saturate the moist air under the same temperature (m_{vs}). It can also be defined as the ratio of partial pressure of water vapor in the air-vapor mixture (p_w) to the partial pressure of water vapor in saturated air at the same temperature (p_{ws}). Relative humidity is expressed using the equation below (Fratoddi et al., 2016):

$$RH = \frac{m_v}{m_{vs}} = \frac{p_w}{p_{ws}} \tag{9.11}$$

where p_{ws} is the partial pressure of water vapor at saturation, kPa.

9.4.4 Saturation Humidity

Saturation humidity (H_s) refers to the capacity of air to hold water at a specific temperature. This water holding capacity increases with increasing temperature. It can also be referred to as the maximum amount of water that air can contain before the water vapor begins to condense back to the original liquid water at a fixed temperature. The saturation humidity is calculated using the following equation (Exell, 2017):

$$H_s = 0.622 \frac{p_{ws}}{1 - p_{ws}} \tag{9.12}$$

9.4.5 Humid Heat Capacity

The humid heat capacity is a psychrometric property that denotes the heat capacity of moist air. It is usually referred to as the heat energy that is needed to raise the temperature of a kilogram of air along with its associated water vapor by 1°C at a constant pressure. It is estimated using the equations below (Exell, 2017):

$$C_p = C_{p\,dry\,air} + C_{p\,water}\,H \tag{9.13}$$

$$C_p = 1.005 + 1.88H \tag{9.14}$$

where $C_{p\,dry\,air}$ is the specific heat of dry air = 1.005 kJ/kg/K; $C_{p\,water}$ is the specific heat of water vapor = 1.88 kJ/kg/K.

9.4.6 Humid Volume

The humid volume (V_H), which is also referred to as the specific volume of air, is defined as the volume that is taken up by a unit mass of dry air and its associated water vapor. Note that the humidity of air is inversely proportional to its density. It is obtained using the equation below (Exell, 2017):

$$V_H = (0.082T + 22.4)\left(\frac{1}{29} + \frac{H}{18}\right) \tag{9.15}$$

where T is the temperature of air, K.

9.4.7 Psychrometric Chart

A psychrometric chart is a graphical representation of the thermodynamic properties of air. Information about the characteristics of air and water vapor mixture is helpful in a number of applications such as sensible heating, sensible cooling, ventilation, and meteorological processes (Atsonios et al., 2015). The psychrometric chart is vital in determining psychrometric properties without the need to calculate using the previously mentioned equations (Kaur et al., 2015). The psychrometric chart is basically a plot between humidity and temperature of air over which is superimposed a series of

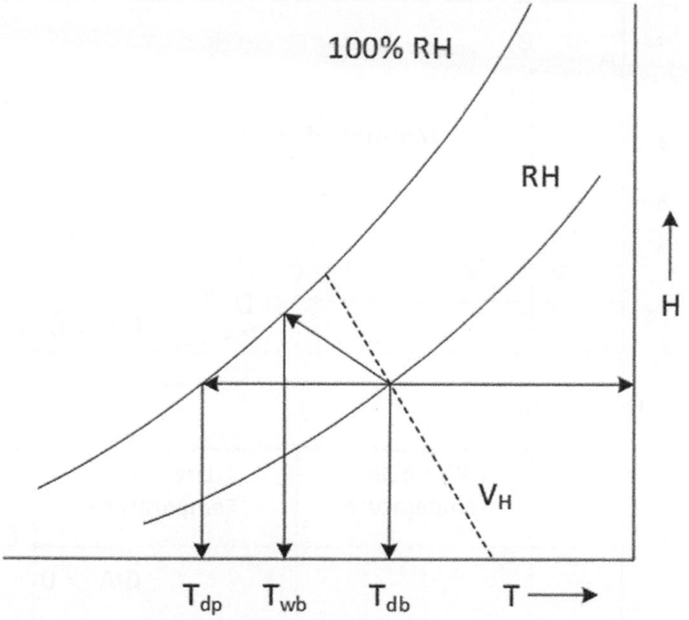

FIGURE 9.1 Example of a psychrometric chart (T_{dp} = dew point temperature, T_{wb} = wet bulb temperature, T_{db} = cry bulb temperature, V_H = humid volume, RH = relative humidity).

curves representing different percentages of relative humidity (Figure 9.1). It is usually illustrated for standard atmosphere pressure (101.325 kPa), which corresponds to the pressure at mean sea level, and it highlights that knowing two properties of air usually aids in uncovering the remaining properties directly using the chart.

9.5 DRYING PROCESSES

9.5.1 Warm-Up Period

The drying processes is usually divided into three parts: the warm-up period, first drying period, and second drying period. From Figure 9.2, you can see that the material that requires drying is first heated from the initial temperature T_p to T_{wb}. The moisture then starts to evaporate from the surface. The mass and heat transfer driving forces are usually non-zero. The process then runs on a drying curve from point A to point B [X_{Ap} to X_A ($\tau = 0$)], and the concentration of water decreases. The drying of water from the surface slows down the heating of the material as the heat from the air is consumed during the process of evaporation. However, the warm-up period should be relatively short and, in some instances, it can be non-catchable.

9.5.2 First Drying Period

The first drying period is the process characterized by the constant-rate period of drying from point B to point C, as seen in Figure 9.2. At point B, the temperature of the

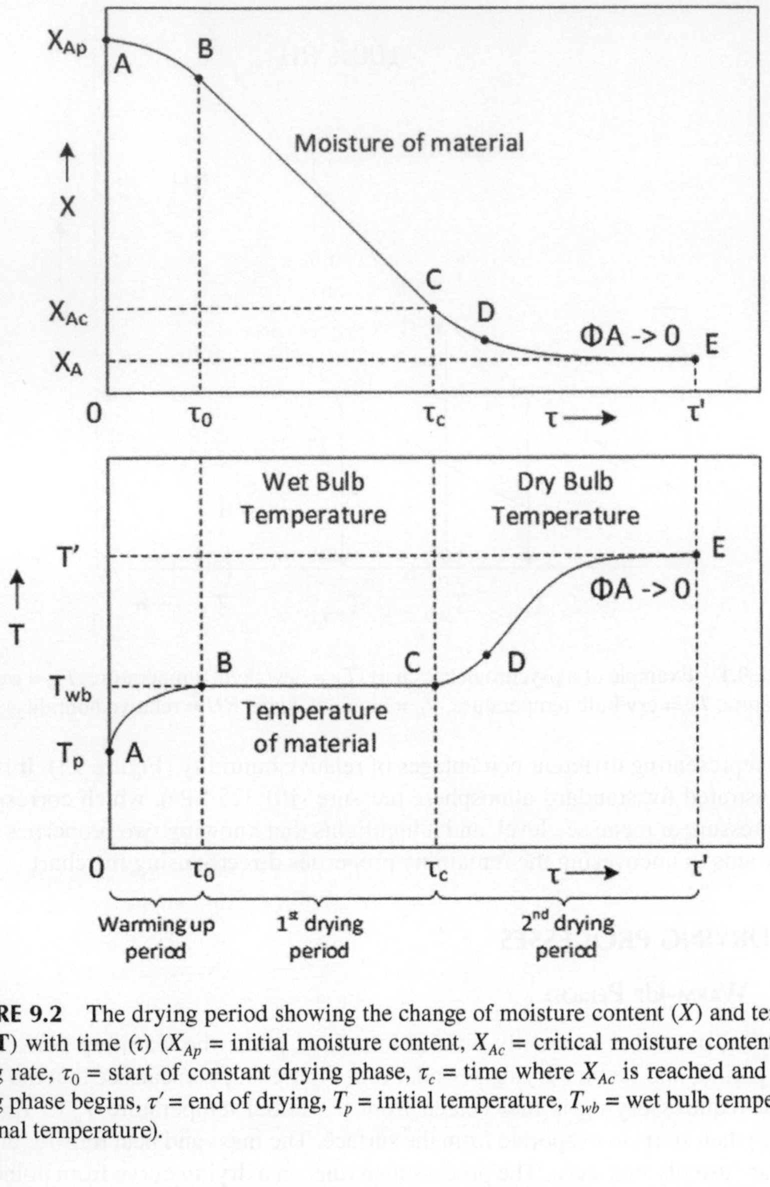

FIGURE 9.2 The drying period showing the change of moisture content (X) and temperature (T) with time (τ) (X_{Ap} = initial moisture content, X_{Ac} = critical moisture content, Φ_A = drying rate, τ_0 = start of constant drying phase, τ_c = time where X_{Ac} is reached and falling drying phase begins, τ' = end of drying, T_p = initial temperature, T_{wb} = wet bulb temperature, T' = final temperature).

material is equal to the wet bulb temperature and usually stays constant because the added heat is consumed via evaporation of free water on the surface of the material. The drying force and drying rate Φ_A are usually constant. The vapor pressure on the material surface during the constant-rate period is usually equal to the pressure of water vapor measured on clean water at temperature T_w. At point C, the moisture content is equal to the critical moisture content X_{Ac}. In the constant-rate period, the rate of drying usually depends on the moisture content, temperature, and mass flow of air. However, it is vital to note that the drying rate is not a function of actual moisture content and bed depth.

9.5.3 Second Drying Period

In Figure 9.2, the water starts percolating at point C; the material's surface then dries up and the first material particles get air contact. The rate of drying starts decreasing, and the temperature of the material starts to rise at point D, which causes a decrease in mass and heat transfer driving forces. The path of the evaporated water vapor is more complicated as the resistance is larger. At point E, the temperature of the materials is measured to be nearly equivalent to that of the air temperature, while the moisture content is equivalent to the equilibrium moisture content of the surrounding air (designated X_A). The heat and mass transfer stopping point is at Φ_A, and it is impossible under real-life conditions to dry the material further. In theoretical terms, achieving the equilibrium point takes a long time.

9.6 INDUSTRIAL APPLICATIONS OF DRYING

9.6.1 Food Drying

One of the major applications of drying is found in food preservation. Food drying entails dehydrating food materials to inhibit the growth of yeasts, molds, and bacteria (Kumar et al., 2015). Drying is usually the last processing stage in the manufacturing of dehydrated food products. For this reason, it brings a reduction in moisture but may also affect some properties of the food such as texture, color, viability, nutrient retention, flavor. These properties are usually affected during changes in temperature and moisture content, which is experienced by the food materials as they move through the dryer.

In industrial food drying, mechanical drying is mostly used. Mechanical drying encapsulates various types of drying methods including hot air convection, freeze drying, microwaving, vacuum drying, fluidized bed drying, and vacuum-assisted microwave drying. In convective drying, hot air is blown through the material either by cross flow or parallel flow. Convective drying saves a great amount of time compared to traditional solar drying. However, for food that loses its aroma at high temperatures, such as fruits and vegetables, freeze drying is preferred. In freeze drying, the food is flash frozen and then put in a pressure reduction system that causes the water to sublimate directly. For microwave drying, food is dried by using volumetric heating. Microwaves causes vibrations in polar molecules (for example, water) by inducing alternating electrical fields. Since food usually contain 52%–99% water, it can be dried effectively using microwaves. However, when it comes to mass production of dry food, convective dryers are most often used because they are cheaper compared to other drying methods such as freeze drying, vacuum drying, and microwaving. Even so, the products that are dried by these mechanical methods are often poor in color, flavor, texture, and rehydration qualities (Kumar et al., 2015).

9.6.2 Chemical and Medical Industries

In the chemical industry, solids drying is utilized to remove water from detergents, catalysts, and polymers as well as other fine pellets, granules, and powders.

Even though convection using fluidized bed is a common way of drying solids, chemical engineers have also considered vacuum drying. For instance, vacuum drying was coupled with a desiccation process called zeodration, a technology that comes from a company called Zedrys. In zeodration, zeolite, a proprietary compound, is used to adsorb water molecules. The zeolite is made of crystalized clay formed from volcanic ash via natural means. The adsorbed water blocks the passage of the products through the zeolite layer, thus allowing complete drying without any product loss. The process is usually started by loading the material in a vacuum chamber, where water evaporates readily under the reduced pressure. The adsorption process is usually exothermic; thus, the reactors must be cooled. The captured heat returns to the vacuum chamber where the lower pressure increases evaporation. The zeolite catalyst is regenerated with hot water or steam as heat dislodges the water molecules from the zeolite pores. This technology has been applied to blood platelets to increase their shelf life. For example, in one study, platelets were dried at room temperature under vacuum pressure (145 mbar). Rehydration of the platelets showed an 85% cell viability, indicating that zeodration could be used as an alternative drying method (Donnet et al., 2015).

Liquid drying is also applied in the chemical industry, where it is mainly coupled with spray drying to facilitate powder formation. Spray drying is a mass transfer method, so the concentration gradient across both the surface area and interface of a gas-liquid is a vital parameter. Using dry air and dispersing liquid into fine droplets maximize the concentration gradient, and this enables rapid drying and powder formation with uniform characteristics.

9.6.3 Drying of Textile Materials

Textile materials can be dried at several stages in the production sequence. In the wet-processing industry, the drying machine is usually used after fabric dewatering. In the textile finishing unit, the dryer is used for drying knit and woven fabrics. The objectives of the drying machine include drying the fabric with the help of steam, controlling shrinkage, preparation for the next subsequent process, as well as drying tubular and open-width fabric without tension. The main parts of the drying machine include a heating chamber, blower, synthetic blanket as conveyor, folder, and exhaust fan. Drying is essential in the textile industry because it is used in eliminating or reducing water from fibers, fabrics, and yarns following the wet processes. The process of drying happens when the liquid is vaporized from the textile product by applying heat. The drying methods employed include convection, infrared radiation, radio-frequency, and contact drying (Wei et al., 2017).

9.6.4 Drying in Mineral Processing

In most cases, wet beneficiation of mineral ores requires the removal of water when water is in excess, and the ores are contaminated. The water removal process occurs before further processing of concentrated ores. Mineral ores and concentrates are usually transported over vast distances by sea or ground transport. Dewatering and drying are of economic importance; mostly they entail selecting dryers that capitalize on superheated steam drying or microwave drying. The end product for most minerals does not require water content, which mandates the need for drying (Wu et al., 2010).

9.6.5 Sludge Drying

Sludge is the residue from refining and industrial processes and is mostly semisolid. It is mainly separated solids that are suspended in liquid and contains large quantities of interstitial water between the particles. As such, the sludge needs to be dried to reduce its moisture content and volume. Dewatering reduces the water content to concentrate the solid content, and it is usually done using mechanical means like drying beds or filter presses. Drying then ensues to evaporate the remaining water. The volume reduction of the sludge makes it easier to store, transport, package, and sell. The end result has many applications, such as for biofuel production, fertilizers, and soil conditioners (Bennamoun et al., 2013).

9.7 DRYING TECHNOLOGY

9.7.1 Stand-Alone Drying Methods

9.7.1.1 Horizontal and Vertical Dryers

There are two types of air flow mechanisms in dryers: horizontal and vertical dryers. These refer to the direction of air flow through the dryers. In horizontal dryers, air is moved from the back side of the dryer to the front side in a horizontal direction. Therefore, the heating element and fan are located at the sides of the dryer. A major advantage of horizontal dryers is uniform heating. In drying of food, different food can be dried at once since moisture dripping from the food is eliminated. In vertical dryers, air is moved from the bottom of the dryer toward the top. The heating element and fan are located at the base of the dryer. A major disadvantage of vertical dryers is that it causes mixing of the flavor of different foods because liquids can drip onto heating elements (Baker, 1992).

9.7.1.2 Solar Drying

Solar drying, in the form of open sun drying, was one of the first drying techniques used in ancient times for drying of agricultural products. Solar drying has disadvantages, however: product spillage in rain and wind; contamination by moisture and dust; product decomposition; loss to birds, animals, and insects; and so on. To overcome these problems, different solar drying systems were developed. Solar drying is broadly classified into two groups: active (forced circulation) and passive (natural circulation) systems. Solar drying is further classified based on mode of operation, namely, direct and indirect types. Direct solar drying involves placing the materials in an enclosure with a transparent cover, and the materials are heated by absorbing sunlight. Indirect solar drying uses air, preheated in a solar collector, as a medium to dry the materials in the drying chamber. Apart from the two classifications, solar dryers are also categorized as integral or distributed systems based on the placement of their solar collectors. If the solar collector shares the same building as the drying chamber, it is an integral system. If the solar collector and drying chamber are separate units, it is a distributed system (Akarslan, 2012).

9.7.1.3 Intermittent Drying

Intermittent drying is a drying method in which the drying conditions are altered with time. Intermittent drying is considered one of the most energy-efficient drying methods (Chua et al., 2003; Kowalski and Pawłowski, 2011). There are various ways to perform intermittent drying, such as varying drying air temperature, air flow rate, pressure, humidity, and type of heat input. To generate intermittency in heat or energy input, different approaches to vary the drying temperature have been studied, including on-off (Chin and Law, 2010), step-up and step-down (Chua et al., 2002), square (Chua et al., 2002; Ho et al., 2002), sawtooth and sinusoidal (Ho et al., 2002) as well as cosine systems (Chua et al., 2002).

Intermittent drying reduces both the effective drying time and drying air utilization; therefore, it uses less energy compared to continuous drying (Putranto et al., 2011). Intermittent drying also introduces tempering periods, which refers to the nonheating durations where the moisture content gradient, created during drying, can diminish. By lowering the moisture content gradient, tempering allows moisture to propagate from the core to the outer layers of the food products in a phenomenon called moisture leveling or moisture uniformity (Kunze and Choudhury, 1972). Longer tempering periods lead to increased moisture leveling and higher initial moisture removal in subsequent active drying time (Jumah et al., 2007).

9.7.1.4 Spray Drying

Spray drying is used for liquid feedstock drying, where the liquid feedstock is sprayed into a hot drying medium and transformed from a liquid state to a dried product. The feedstock can be a solution, emulsion, or suspension, and the dried product can be powder, granule, or agglomerate. The final form for the dried product depends on the physical and chemical characteristics of the feedstock, the desired characteristics of the end product, and the design of the spray dryer (Patel et al., 2009).

9.7.1.5 Oven Drying

Oven drying is one of the simplest drying techniques used in industry, but it has a lower energy efficiency than other drying methods. Compared to a dehydrator, oven drying of food takes twice as long to reach the same moisture content. Various types of oven dryers have different features. For instance, a vacuum oven dryer carries out drying at a reduced pressure, whereas a microwave oven dryer utilizes microwaves for heating. Compared with vacuum ovens, microwave ovens feature higher precision and flexibility for in-process adjustments of the moisture content in food. Microwave oven drying also has the shortest drying time, followed by vacuum oven drying (Arslan and Özcan, 2010).

9.7.1.6 Superheated Steam Drying

Superheated steam drying (SSD) involves passing dry superheated steam at 250°C–300°C through the drying chamber, where the sensible heat of the steam is used to evaporate moisture from the feedstock. Superheated steam (SS) is generated by continuously adding heat energy to steam to increase its temperature above the saturation temperature at a specific pressure. It is not essential to vent the evaporated water from the product before the pressure in the drying chamber reaches a certain limit. After that, excess steam can be vented. Despite the high temperature of SS, the product temperature does not reach the temperature of the SS (Sagar and Kumar, 2010). Table 9.2 lists the key advantages and disadvantages of SSD.

9.7.2 Hybrid Drying Systems

The diversity of food products has demanded the use of different types of dryers. But sometimes a single drying method is insufficient for higher-quality products, unsuitable for heat-sensitive products, or unable to save on cost. However, hybrid drying systems can be used to offset these limitations. Hybrid drying systems integrate

TABLE 9.2
Advantages and Disadvantages of Superheated Steam Drying (SSD)

Advantages	Disadvantages
1. No combustion or oxidation reactions can occur in SSD, making the system free from fire or explosion hazard	1. SSD system is more complex than air dryer. No leakage is permitted. No infiltration of air is permitted during feeding and discharging steps. Start-up and shutdown require complex operation
2. High temperature of superheated steam (SS) allows greater drying rates in both constant and falling rate periods	2. When feed enters at ambient temperature, condensation occurs before evaporation begins, and this prolongs residence time in the dryer by 10%–15%
3. Many feedstocks that form "case-hardened skin" due to rapid drying do not form this water-impermeable skin in SSD	3. SSD cannot dry products that can deform or be damaged at the saturation temperature of steam at the dryer operating pressure
4. SSD minimizes fire and explosion risks when drying products containing toxic or valuable organic solvents, although condensation of off-streams can occur in relatively smaller condensers	4. SSD cannot dry products that require oxidation reactions (for example, browning of food)
5. SSD allows sterilization, pasteurization, and deodorization of food products	5. If the heat energy in the generated steam is not recovered, the energy savings of SSD do not exist
	6. Steam cleaning may be difficult depending on the chemical composition of the condensate
	7. The high capital cost of SSD makes it a justifiable option only for continuous operation of very large amounts of feed
	8. Currently, there is limited experience in using SSD for smaller range of products

Source: Mujumdar, A. S., Chapter 20—Superheated steam drying, in *Handbook of Industrial Drying*, edited by Mujumdar, A. S., CRC Press, Boca Raton, FL, pp. 421–432, 2014.

more than one type of drying method and/or mode of heat transfer to minimize product quality degradation and improve energy efficiency (Chou and Chua, 2001).

9.7.2.1 Hybrid Solar Drying

Hybrid solar dryers are most commonly used because they are more energy efficient and cost effective compared to conventional solar dryers. Hybrid solar dryers combine solar dryers and other drying methods, such as heat pumps, photovoltaic systems, thermal energy storage, and others. Ayyappan et al. investigated the performance of several sensible heat storage materials (concrete, sand, and rock bed) in a passive solar greenhouse dryer. An optimum thickness of 4″ was used, and the concrete, sand, and rock bed dried coconuts from 52% (w.b.) to 7% (w.b.) in 78 h, 66 h, and 53 h, respectively. In comparison, open sun drying of the coconuts took 174 h to reach the same moisture content. The dryer efficiencies using concrete, sand, and rock bed were observed to be 9.5%, 11%, and 11.7%, respectively (Ayyappan et al., 2016).

9.7.2.2 Heat Pump Drying

Heat pump dryers (HPDs) improve energy efficiency when integrated with other drying methods (Chua and Chou, 2005). A heat pump is a device that transfers heat energy from a cold region (evaporator) to a hot region (condenser) using mechanical energy, as in a refrigerator. In the same sense, HPDs can recover heat energy from exhaust gas (evaporator end) and convert it into sensible heat to heat the drying gas (condenser end) (Colak and Hepbasli, 2009). This implies that any dryer utilizing convective heat can be fitted with an appropriately sized heat pump (Chua et al., 2010). The energy efficiency of HPDs can be seen from their higher specific moisture extraction ratio (SMER) values as compared to other drying methods (see Table 9.3). Table 9.4 lists the key advantages and disadvantages of HPDs.

9.7.2.3 Ultrasound-Assisted Drying

Due to the high heat capacity of water, drying is usually a long and energy-intensive process. The amount of energy used causes damage to the product quality. The quality and characteristics of food products can be preserved by using high power–low

TABLE 9.3
Comparison of Heat Pump Drying and Other Drying Methods

Parameter	Convective Dryer	Vacuum Dryer	Heat Pump Dryer
Dryer efficiency (%)	35–40	≤70	Up to 95
Temperature range (°C)	40–90	30–60	10–65
RH range (%)	Variable	Low	10–65
SMER (kg water/kW h)	0.12–1.28	0.72–1.2	1.0–4.0
Initial cost	Low	High	Medium
Operational cost	High	Very high	Low

Source: Perera, C. O. and Rahman, M. S., *Trends Food Sci. Technol.*, 8, 75–79, 1997.
Abbreviations: RH, relative humidity; SMER, specific moisture extraction ratio.

TABLE 9.4
Advantages and Disadvantages of Heat Pump Dryers

Advantages	Disadvantages
1. High SMER value as heat energy can be recovered from the exhaust gas	1. The chlorofluorocarbons (CFCs) used as the refrigerants are not environmentally friendly
2. Wide range of drying temperature (−20°C–100°C with auxiliary heating) and RH (15%–80% with humidifier)	2. Periodic maintenance of equipment is necessary and refrigerant have to be refilled
3. Suitable for drying food products at low temperatures, thereby enhancing product quality	3. Increased capital cost
4. Greater energy efficiency for drying heat-sensitive food products compared to energy-consuming methods like freeze drying	4. Limited drying temperature
	5. Complex process control and design

Source: Chou, S. K. and Chua, K. J., *Trends Food Sci. Technol.*, 359–369, 2001; Chou, S. K. and Chua, K. J., Heat pump drying systems, in *Handbook of Industrial Drying*, edited by Mujumdar, A. S., CRC Press, Boca Raton, FL, pp. 1103–1132, 2006.
Abbreviations: RH, relative humidity; SMER, specific moisture extraction ratio.

frequency ultrasound drying. Ultrasonic frequency employed in high-power ultrasound operations is normally about 20 kHz because the vibrational energy of higher frequencies tend to be absorbed by the medium, causing a reduction of available energy reaching the solid-fluid interface. The mechanism of ultrasound-assisted drying utilizes sound pressure to cause a series of rapid, successive compressions and rarefactions in the feed. The resulting ultrasonic vibrations helped to push water toward the surface of the feed, where the water can be evaporated by hot air. Treatment using ultrasound-assisted drying improves the drying process by increasing the heat and mass transfer (Chemat and Khan, 2011). Providing an ultrasonic field may stimulate diffusion at the interface between a fluid and suspended solid, and heat transfer is enhanced by 30%–60% depending on the ultrasound intensity (Gallego-Juárez, 1998).

In spray dryers, ultrasonic nozzle systems are used to improve the efficiency of liquid atomization. Liquids are pumped through the orifice at high pressure and velocity, and increasing the intensity of piezoelectric crystals at the nozzles improves the liquid atomization (Tatar Turan et al., 2015). In solid-liquid systems, ultrasound has been used to accelerate mass transfer of food products, such as in the preservation of fruits like apples in sugar solution, osmotic dehydration of apples, and brining of cheese and meat. But some researchers have also found that ultrasound did not show significant effects in drying of food products (Cárcel et al., 2007). More research needs to be done to properly evaluate the efficacy of ultrasound-assisted drying.

9.7.2.4 Microwave-Assisted Drying

Microwaves are electromagnetic waves with frequencies ranging between 300 MHz to 300 GHz. Domestic microwave appliances commonly use a frequency of 2.45 GHz, whereas industrial microwave applications utilize frequencies of 915 MHz and 2.45 GHz (Dibben, 2001). Microwave drying has the benefit of fast drying rates;

improved product quality; and selective heating of interior, moisture-containing regions of feed with minimal effect on the exterior regions. Microwave drying is particularly useful when applied in the falling rate period. During the falling rate period, diffusion of water from the interior to the exterior of feed is rate-limiting, and this often leads to shrinkage of feed and a decrease in surface moisture content. When microwaves are used, water in the interior regions is heated and vapors are formed. This generates an internal pressure gradient that forces the water out of the feed. In this way, feed shrinkage is kept to a minimum. One of the major disadvantages of microwave drying is the difficulty in controlling the feed temperature compared to hot air drying, where the feed temperature never goes above the hot air temperature. Microwaves can cause high temperatures along the corner or edges of food products, resulting in scorching and production of off-flavors, particularly during final drying processes. For large-scale drying, the penetration depth of microwaves at 2.45 GHz is limited, but the penetration depth can be improved by using radio frequency heating at a variable range of 10–300 MHz. In the case of some food products, rapid mass transport of water may alter the food structure in a process called puffing, which may be desirable or unwanted depending on the final food product. Therefore, it is a better option to implement microwave drying with other drying methods in order to enhance drying efficiency and product quality compared to using stand-alone drying methods (Zhang et al., 2006).

9.8 CONCLUSION

Drying provides an effective preservation mechanism in many industries to increase the shelf life or storage duration of their products. Drying is commonly the last stage of the process to prevent changing the product quality. The high heat capacity of water results in a high energy requirement for the evaporation of water molecules from products. The large amount of energy used during the drying process could decrease the product quality. Therefore, hybrid drying technologies are being developed that have higher drying efficiency and enable closer monitoring of product quality. Some of the new hybrid technologies being developed commercially to overcome limitations of drying are ultrasound-assisted and microwave-assisted drying. The quality and characteristics of final products can be preserved by using high power–low frequency ultrasound-assisted drying. Microwave-assisted drying provides rapid and uniform heat distribution, leading to higher drying efficiency and lower energy consumption. Despite these results, more research is still required to develop hybrid drying technologies as well as increasing both the drying and energy efficiency of existing drying methods.

REFERENCES

Abass, A. B., G. Ndunguru, P. Mamiro et al. 2014. "Post-harvest food losses in a maize-based farming system of semi-arid savannah area of Tanzania." *Journal of Stored Products Research* 57: 49–57.

Akarslan, F. 2012. "Solar-Energy Drying Systems." In *Modeling and Optimization of Renewable Energy Systems*, edited by Arzu Sencan, pp. 1–20. Rijeka, Croatia: InTech.

Arslan, D., and M. M. Özcan. 2010. "Study the effect of sun, oven and microwave drying on quality of onion slices." *LWT-Food Science and Technology* 43(7): 1121–1127.

Atsonios, K., I. Violidakis, M. Agraniotis et al. 2015. "Thermodynamic analysis and comparison of retrofitting pre-drying concepts at existing lignite power plants." *Applied Thermal Engineering* 74: 165–173.

Ayyappan, S., K. Mayilsamy, and V. Sreenarayanan. 2016. "Performance improvement studies in a solar greenhouse drier using sensible heat storage materials." *Heat and Mass Transfer* 52(3): 459–467.

Baker, C. G. J. 1997. *Industrial Drying of Foods*. 1st ed. London, UK: Chapman & Hall.

Beltrán-Prieto, J. C., L. A. Beltrán-Prieto, and L. H. B. S. Nguyen. 2016. "Estimation of psychrometric parameters of vapor water mixtures in air." *Computer Applications in Engineering Education* 24(1): 39–43.

Bennamoun, L., P. Arlabosse, and A. Léonard. 2013. "Review on fundamental aspect of application of drying process to wastewater sludge." *Renewable and Sustainable Energy Reviews* 28: 29–43.

Cárcel, J. A., J. Benedito, J. Bon, and A. Mulet. 2007. "High intensity ultrasound effects on meat brining." *Meat Science* 76(4): 611–619.

Chemat, F., and M. K. Khan. 2011. "Applications of ultrasound in food technology: Processing, preservation and extraction." *Ultrasonics Sonochemistry* 18(4): 813–835.

Chin, S. K., and C. L. Law. 2010. "Product quality and drying characteristics of intermittent heat pump drying of Ganoderma tsugae Murrill." *Drying Technology* 28(12): 1457–1465.

Chou, S. K., and K. J. Chua. 2001. "New hybrid drying technologies for heat sensitive foodstuffs." *Trends in Food Science and Technology* 12(10): 359–369.

Chou, S. K., and K. J. Chua. 2006. "Heat pump drying systems." In *Handbook of Industrial Drying*, edited by A. S. Mujumdar, pp. 1103–1132. Boca Raton, FL: CRC Press.

Chua, K., and S. Chou. 2005. "A modular approach to study the performance of a two-stage heat pump system for drying." *Applied Thermal Engineering* 25(8–9): 1363–1379.

Chua, K., S. Chou, M. Hawlader, A. Mujumdar, and J. Ho. 2002. "PH–Postharvest technology: Modelling the moisture and temperature distribution within an agricultural product undergoing time-varying drying schemes." *Biosystems Engineering* 81(1): 99–111.

Chua, K., A. Mujumdar, and S. Chou. 2003. "Intermittent drying of bioproducts—An overview." *Bioresource Technology* 90(3): 285–295.

Chua, K. J., S. K. Chou, and W. M. Yang. 2010. "Advances in heat pump systems: A review." *Applied Energy* 87(12): 3611–3624.

Colak, N., and A. Hepbasli. 2009. "A review of heat pump drying: Part 1—Systems, models and studies." *Energy Conversion and Management* 50(9): 2180–2186.

Dibben, D. 2001. "Electromagnetics: Fundamental aspects and numerical modeling." In *Handbook of Microwave Technology for Food Application*, edited by A. K. Datta and R. C. Anantheswaran, pp. 1–31. New York: CRC Press.

Donnet, T., C. Ravanat, A. Eckly et al. 2015. "Dehydration of blood platelets by zeodration: In vitro characterization and hemostatic properties in vivo." *Transfusion* 55(9): 2207–2218.

Engineering Toolbox. *Dry Bulb, Wet Bulb and Dew Point Temperatures* 2004 [cited August 3, 2018]. Available from https://www.engineeringtoolbox.com/dry-wet-bulb-dew-point-air-d_682.html.

Exell, R. 2017. "Basic design theory for a simple solar rice dryer." *International Energy Journal* 1(2): 1–14.

Fratoddi, I., A. Bearzotti, I. Venditti, C. Cametti, and M. Russo. 2016. "Role of nanostructured polymers on the improvement of electrical response-based relative humidity sensors." *Sensors and Actuators B: Chemical* 225: 96–108.

Gallego-Juárez, J. 1998. "Some applications of air-borne power ultrasound to food processing." In *Ultrasound in Food Processing*, pp. 127–143, London, UK: Chapman & Hall.
Ho, J., S. Chou, K. Chua, A. Mujumdar, and M. Hawlader. 2002. "Analytical study of cyclic temperature drying: effect on drying kinetics and product quality." *Journal of Food Engineering* 51(1): 65–75.
Jumah, R., E. Al-Kteimat, A. Al-Hamad, and E. Telfah. 2007. "Constant and intermittent drying characteristics of olive cake." *Drying Technology* 25(9): 1421–1426.
Kaur, B. P., V. S. Sharanagat, and P. K. Nema. 2015. "Fundamentals of Drying." In *Drying Technologies for Food: Fundamentals and Applications*, edited by P. K. Nema, pp. 1–22. New Delhi, India: New India Publishing Agency.
Kowalski, S., and A. Pawłowski. 2011. "Energy consumption and quality aspect by intermittent drying." *Chemical Engineering and Processing: Process Intensification* 50(4): 384–390.
Kumar, D., and P. Kalita. 2017. "Reducing postharvest losses during storage of grain crops to strengthen food security in developing countries." *Foods* 6(1): 8.
Kumar, Y., S. Tiwari, and S. A. Belorkar. 2015. "Drying: An excellent method for food preservation." *International Journal of Engineering Studies and Technical Approach* 1(8): 1–17.
Kunze, O., and M. Choudhury. 1972. "Moisture adsorption related to the tensile strength of rice." *Cereal Chemistry* 49: 684–696.
Mujumdar, A. S. 2014. "Chapter 20–Superheated steam drying." In *Handbook of Industrial Drying*, edited by A. S. Mujumdar, pp. 421–432. Boca Raton, FL: CRC Press.
Patel, R., M. Patel, and A. Suthar. 2009. "Spray drying technology: An overview." *Indian Journal of Science and Technology* 2(10): 44–47.
Perera, C. O., and M. S. Rahman. 1997. "Heat pump dehumidifier drying of food." *Trends in Food Science and Technology* 8(3): 75–79.
Prajapati, V. K., P. K. Nema, and S. S. Rathore. 2011. "Effect of pretreatment and drying methods on quality of value-added dried aonla (Emblica officinalis Gaertn) shreds." *Journal of Food Science and Technology* 48(1): 45–52.
Putranto, A., X. D. Chen, S. Devahastin, Z. Xiao, and P. A. Webley. 2011. "Application of the reaction engineering approach (REA) for modeling intermittent drying under time-varying humidity and temperature." *Chemical Engineering Science* 66(10): 2149–2156.
Sagar, V., and P. S. Kumar. 2010. "Recent advances in drying and dehydration of fruits and vegetables: A review." *Journal of Food Science and Technology* 47(1): 15–26.
Smith, P. G. 2011. "An introduction to food process engineering." In *Introduction to Food Process Engineering*, pp. 1–3. New York: Springer.
Tatar Turan, F., A. Cengiz, and T. Kahyaoglu. 2015. "Evaluation of ultrasonic nozzle with spray-drying as a novel method for the microencapsulation of blueberry's bioactive compounds." *Innovative Food Science and Emerging Technologies* 32: 136–145.
Vega, R. E., P. Zúñiga, J. Thibault, R. Blasco, and P. I. Álvarez. 2016. "Indirect drying of copper concentrate in a rotating-coil dryer." *Journal of Fluid Flow, Heat and Mass Transfer* 3(1): 62–72.
Wei, Y., J. Hua, and X. Ding. 2017. "A mathematical model for simulating heat and moisture transfer within porous cotton fabric drying inside the domestic air-vented drum dryer." *The Journal of the Textile Institute* 108(6): 1074–1084.
Wojdyło, A., A. Figiel, K. Lech, P. Nowicka, and J. Oszmiański. 2014. "Effect of convective and vacuum–microwave drying on the bioactive compounds, color, and antioxidant capacity of sour cherries." *Food and Bioprocess Technology* 7(3): 829–841.
Wu, Z., Y. Hu, D. Lee, A. Mujumdar, and Z. Li. 2010. "Dewatering and drying in mineral processing industry: Potential for innovation." *Drying Technology* 28(7): 834–842.

Zhang, M., H. Chen, A. S. Mujumdar et al. 2017. "Recent developments in high-quality drying of vegetables, fruits, and aquatic products." *Critical Reviews in Food Science and Nutrition* 57(6): 1239–1255.

Zhang, M., J. Tang, A. S. Mujumdar, and S. Wang. 2006. "Trends in microwave-related drying of fruits and vegetables." *Trends in Food Science & Technology* 17(10): 524–534.

Zhang, S., L. Zhou, B. Ling, and S. Wang. 2016. "Dielectric properties of peanut kernels associated with microwave and radio frequency drying." *Biosystems Engineering* 145: 108–117.

Zorya, S., N. Morgan, L. Diaz Rios et al. 2011. Missing food: The case of postharvest grain losses in sub-Saharan Africa. Washington, DC: World Bank.

Index

Printed and bound by CPI Group (UK) Ltd, Croydon, CR0 4YY
24/10/2024
01778307-0003